2019/9/7

MORE THAN

150

Questions

PART

5

2019

Vibration Analysis Certification

Exam Preparation Package

Certified Vibration Analyst Category I

FAULT ANALYSIS AND CORRECTION

Ali M. Al-Shurafa

Title	Vibration Analysis Certification Exam Preparation Package Certified Vibration Analyst Category I: Fault Analysis and Correction	**Author**	Ali M. Al-Shurafa
Subtitle	ISO 18436-2 CVA Level 1: Part 5	**Series**	CAT I PREP I SERIES PRACTICE TESTS FOR CVA
ISBN-10	1-64415-008-5	**ISBN-13**	978-1-64415-008-5
Publisher	Prep Certify	**Ordering**	www.prepcertify.com

Disclaimers

Although every precaution has been taken to verify the accuracy of the information contained herein, the author and publisher assume no responsibility for any errors. No liability is assumed for damages that may result from the use of this package. The questions covered in this booklet are related to the standard body of knowledge according to ISO standard 18436-2 Second Edition 2014.

Comments can be sent to info@prepcertify.com

Notes

TABLE OF CONTENTS

Notes

CAT I PREP I AND YOUR CERTIFICATION

This book of the series might be the most wanted, exciting and challenging part because of the topics it covers. Fault Analysis and Corrective Action are what most vibration analysts need to deliver as a service to their client. Though other parts of the Body of Knowledge (like Signal Processing and Equipment Knowledge) are essential, the client considers them as preparation details, not what he/she wants to hear from the analyst. Part 5 (Fault Analysis and Corrective Action) addresses basic machinery faults an analyst is expected to identify and the actions required by the operator, repair person, vendor etc to resolve these faults.

Diagnosis can be understood, in simple terms, as the identification of reasons behind the high vibration. Diagnosis searches for interesting symptoms and link them to the possible faults. Remember that fault diagnosis is always based on incomplete information and there is always a risk of an incorrect call. For this reason, a good analyst tries to support her/his diagnosis by multiple symptoms, facts, inputs etc.

Figure 1 illustrates the basic workflow of the vibration analysis job. It states that if the vibration condition is acceptable, do not go through the fault diagnosis, no need for that. A basic rule is that if the vibration amplitudes are higher than the established limit, vibration data analysis is required and the focus is on the frequency and phase. Amplitude is foundational in the condition evaluation step. Refer to Figure 2.

In the fault diagnosis, plot interpretation is a fundamental skill. So, the analyst needs be comfortable with spectra and waveforms. For instance, you need to be able to make unit conversion of the frequency axis in from cmp to Hz and vice versa. Normalizing the frequencies in the spectrum to the running speed is very important. The wallcharts or diagnosis tables used to guide new comers to the diagnosis field are based on orders or normalized frequencies.

Refer to the Formula Sheet section of this book for a typical diagnosis table. Diagnosis wallcharts are helpful but with a caution that they are is based on certain experience on certain type of equipment. They are simplified and generalized to make diagnosis easier. But they are not accurate or complete 100%.

Equipment knowledge is critical for fault analysis because many faults are equipment type specific. For example, because of the electromagnetic

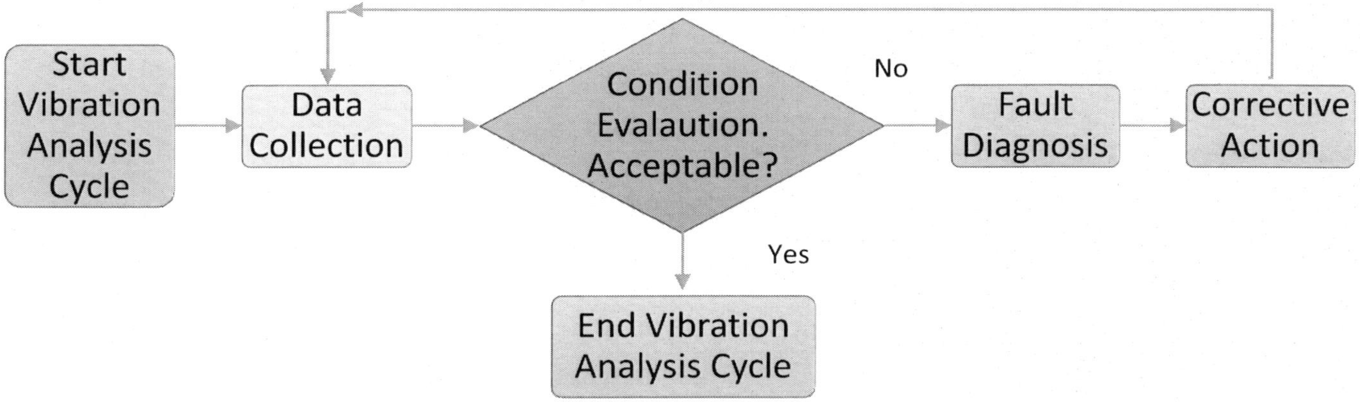

Figure 1 Workflow of a Typical Vibration Analysis Cycle

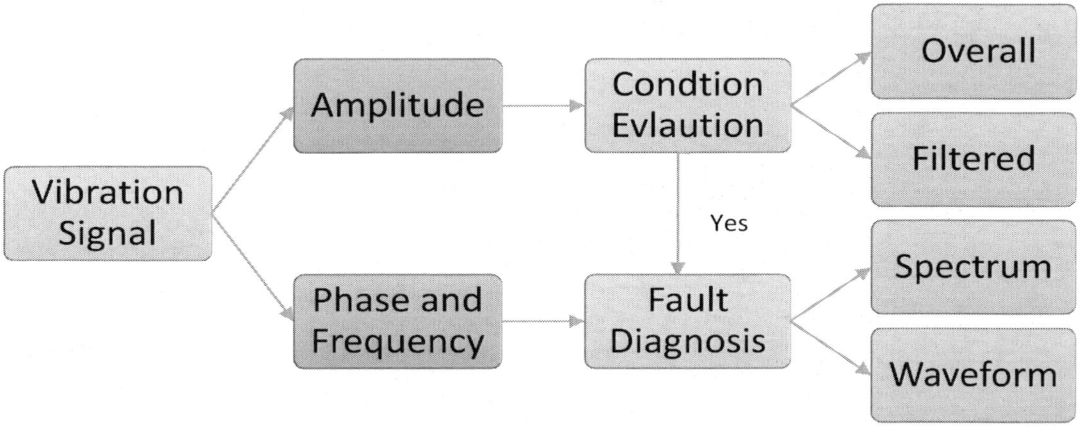

Figure 2 The Use of Frequency and Phase

forces, motors have special faults related to electricity. Pumps can develop high vibration (cavitation) associated with the liquid properties they handle. These mentioned faults are not encountered by gearboxes. Part 5 of this series covers Equipment Knowledge tailored to the vibration certification program.

Following are some tips for better and easier fault analysis using vibration data:

1. Identify the type of data available. For example, ask yourself is it a filtered or unfiltered waveform?

2. Recognize the x-axis units and be ready for some conversions. This is applicable for waveforms and spectra.

3. Read the notes on the plots like the name of the point, the cursor value, additional remarks etc.

4. Pay attention to special patterns in the plot. For example, see the location of the major peaks in a spectrum with respect to the spectrum frequency span. Another example is the magnitude in a waveform if it is stable or not.

5. Perform basic calculations like converting frequency units from cpm to order or vice versa. Calculate fault frequencies like vane pass, gear mesh

and bearing fault frequencies.

6. Link the available information in the question or the plots to each other. Recognize what is known and what is unknown.

7. Translate the written information to a sketch. This step helps you summarize and visualize the available data and relate them. If you are given a plot (e.g. spectrum) or a drawing of an equipment, use them as a place to collect the information.

8. Use the given table for common faults and apply the process of elimination to generate a short list to further analyze.

In the diagnosis, the accuracy of frequency is very important. When you read a plot for frequency (from a spectrum) or calculate it (from a waveform), avoid excessive approximation. For example, a vibration component at 5.1X has a completely different diagnosis compared to 5.0 X. As a rule of thumb, reasonable rounding or approximation should not be more than 2% if you perform the calculations. For simplicity, you can round/approximate the 4th digit like 3582 rpm to 3580 or 12.56 mm/sec to 12.6.

Fault correction comes after completing the diagnosis and is given as a recommendation. Stating what to correct is usually easy. The implementation part is more difficult. For instance, a diagnosis led to rotor unbalance. So, the recommendation to correct rotor unbalance by balancing. But how exactly to perform rotor balancing on the site? So be mindful for the practicality of the recommendation.

This book contains more than 150 questions many of which are challenging for Vibration Analyst Category I. these questions are identified by a special icon added at the end of the question.

Good luck- Ali M. Al-Shurafa

CAT I PART BOK

Below is the Body of Knowledge as specified by the standard ISO18436-2 for Cat I Part 5 “Fault Analysis and Corrective Action”. For a complete list of topics for Cat I, refer to Part 8 or the Standard.

Topic	Details
Fault Analysis	General fault recognition Recognize fault frequencies for pre- set FFT and simple time waveforms for unbalance, loose- ness, misalignment, bearing noise and damage. Also recognize the terms: resonance and phase
Corrective Action	Basic maintenance action Be aware of simple maintenance actions to rectify or reduce faults e.g. lubrication, alignment

FORMULA SHEET

Frequency Order	Possible Fault
Almost zero	False signal. Sudden large movement. Ski slope.
0.3 - 0.7 X	Cage defect of a rolling element bearing
1X	Many faults. Unbalance. Misalignment
2X	Misalignment.
NX (3 - 8)	Misalignment. Vane or blade pass frequency Looseness.
NX (20-100) N: Integer	Many faults. Vane or blade pass frequency. Rotor bar pass, Gear mesh frequency
NX (3.01-30.1) N: non-integer	Inner race, outer race, ball pass frequency and background excitation.
Fault	Possible Order or other symptoms
Rotor unbalance	1X horizontal and vertical similar amplitudes but can be higher in one direction.
Shaft misalignment	Usually 2X or 3X but can be 1X, radial and possibly axial

Rolling element bearing defect	Non-integer order from 0.25 X- 8.9 X. Usually localized on the bearing of defect. Noisy waveform and spectrum.
Process upset on pump, fan and compressors	Vane or blade pass frequency. Many times with noisy waveform and spectrum.
Slip for an induction motor	Synch speed- Running speed
Synchronous Speed for motors and generators	SN=(2xLF)/Poles

Notes

MORE THAN
150
Questions
PART
5
2019

QUESTION BANK

Ali M. Al-Shurafa

Fault Analysis and Correction

Question 001

In the field of machinery vibration, what is true about fault analysis?

A. It is a process of identifying interesting symptoms from the vibration data to short list the most possible causes.

B. It is a process of collecting repeatable vibration data to reliably indicate the condition of the asset.

C. It is a process of reducing vibration amplitude by for example adding or removing mass to/from the rotor.

D. It is a process of specifying the instrumentation requirements to protect the most critical rotating equipment at a plant.

Fault Analysis and Correction

Question 002

How do equipment owners correct high vibrations on rotating equipment?

A. By operating the equipment according to its design.

B. By reducing the forces and moments on the shaft. These forces and moments may be generated by the shaft or acting on the shaft.

C. By performing good quality rotor balancing at the shop and using laser alignment at the site.

D. All of the above.

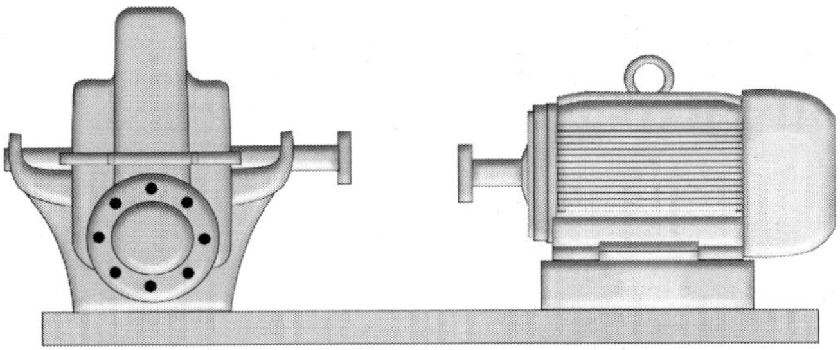

Fault Analysis and Correction

Question 003

Assume there is good quality data for a high machinery vibration due to a mechanical fault. Which of the following is an action that the vibration analyst needs to do to pinpoint the fault?

A. Review historical and current vibration data.

B. Collect additional data such as operating parameters.

C. Talk to the operations and maintenance staff about their observations.

D. All of the above.

Note : ....

....

....

Fault Analysis and Correction

Question 004

In the field of machinery vibration, what is true about fault correction?

A. It is a process of identifying interesting symptoms from the vibration data to identify the most possible faults.

B. It is a process of collecting repeatable vibration data to reliably indicate the condition of the asset.

C. It is a process of reducing vibration amplitude, for example by adding/removing mass to/from the rotor.

D. It is a process of specifying the instrumentation requirements to protect the most critical rotating equipment.

Note : ....

....

....

Fault Analysis and Correction

Question 005

What is true about fault analysis or fault diagnosis when we deal with machinery vibration?

A. It is primarily collecting representative data from the machine such as high resolution spectra.

B. It is primarily giving a clear recommendation to the project team on the acceptance of the machine based on the measured vibration readings.

C. It is primarily a digital processing operation in the vibration analyzer such as filtering, averaging and cross phase calculations.

D. It is primarily identifying the causes of a high vibration problem based on data collected from the machine and its surroundings.

Fault Analysis and Correction

Question 006

What is true about correcting faults detected by vibration analysis on machines in the current industry?

A. The corrective actions are always executed after the trip of the asset.

B. The corrective actions are a pure responsibility of the maintenance team.

C. The corrective actions are needed to be evaluated for the appropriate time and location to be executed.

D. All of the above

Note : ....

....

....

Fault Analysis and Correction

Question 007

The next vibration plot is called waterfall. It is used to compare vibration signatures over a period of time. Which axis refers to the frequency, amplitude and date, respectivey?

	Frequency	Amplitude	Date
A.	1	2	3
B.	1	3	2
C.	2	1	3
D.	3	1	2

Note : ....

....

....

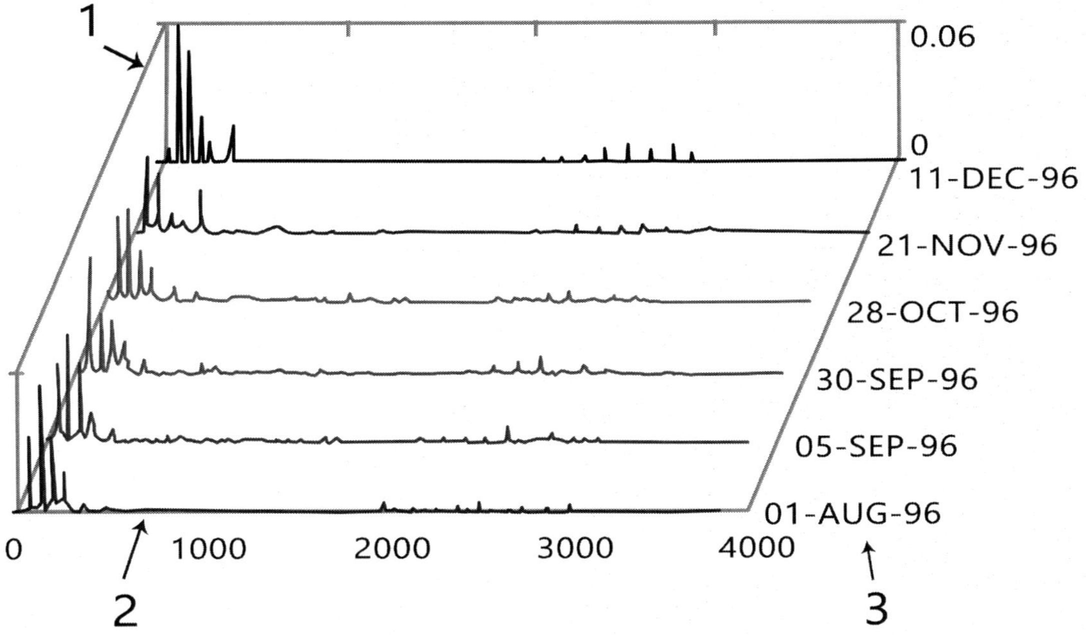
1
0.06
0
11-DEC-96
21-NOV-96
28-OCT-96
30-SEP-96
05-SEP-96
01-AUG-96
0
1000
2000
3000
4000
2
3

Note : ....

....

....

Fault Analysis and Correction

Question 008

Study the provided plots. What is the difference between these vibration waveforms?

A. Units of amplitude and data collection time.

B. Calculation method of the overall amplitude.

C. Range of amplitude and frequency.

D. Plot 2 has a higher resolution.

Note : ....

....

....

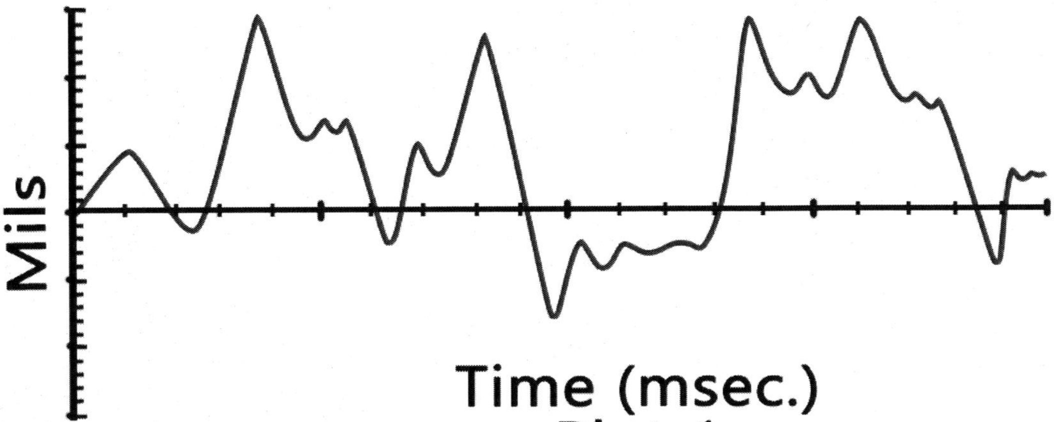

Plot 1

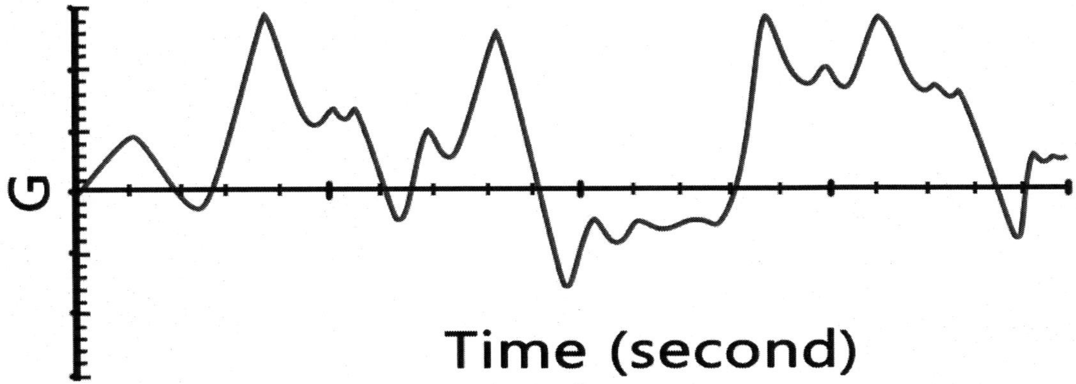

Plot 2

Fault Analysis and Correction

Question 009

Study the provided plots. What is the difference between these vibration waveforms?

A. Units of amplitude, frequency and time.

B. Calculation method of overall amplitude.

C. Plot 2 has longer data acquisition time.

D. Plot 2 has a higher resolution.

Note :

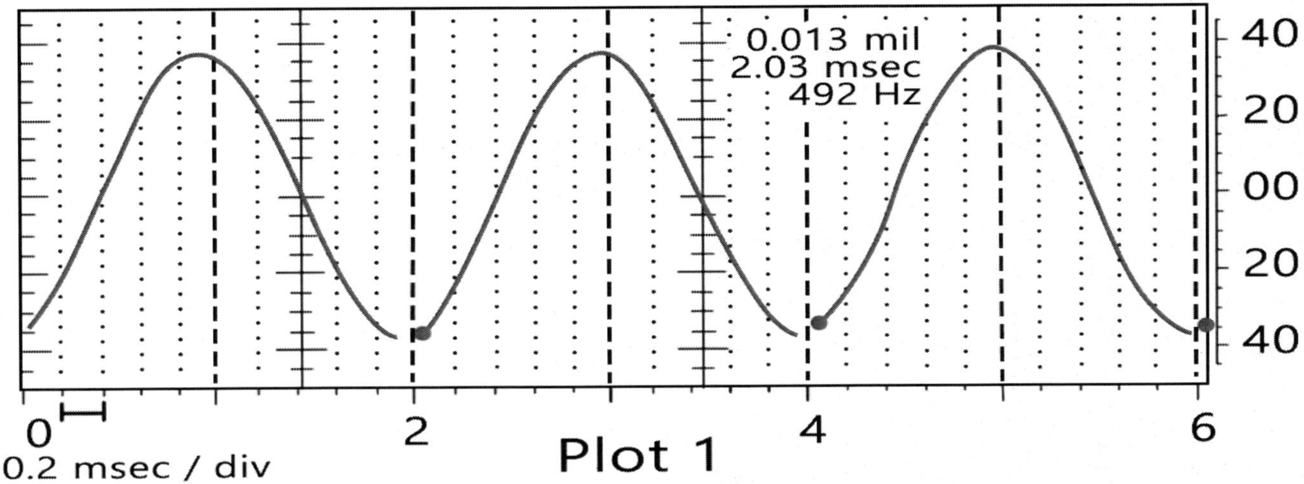
0.013 mil
2.03 msec
492 Hz
40
20
00
20
40
0
2
4
6
0.2 msec / div
Plot 1

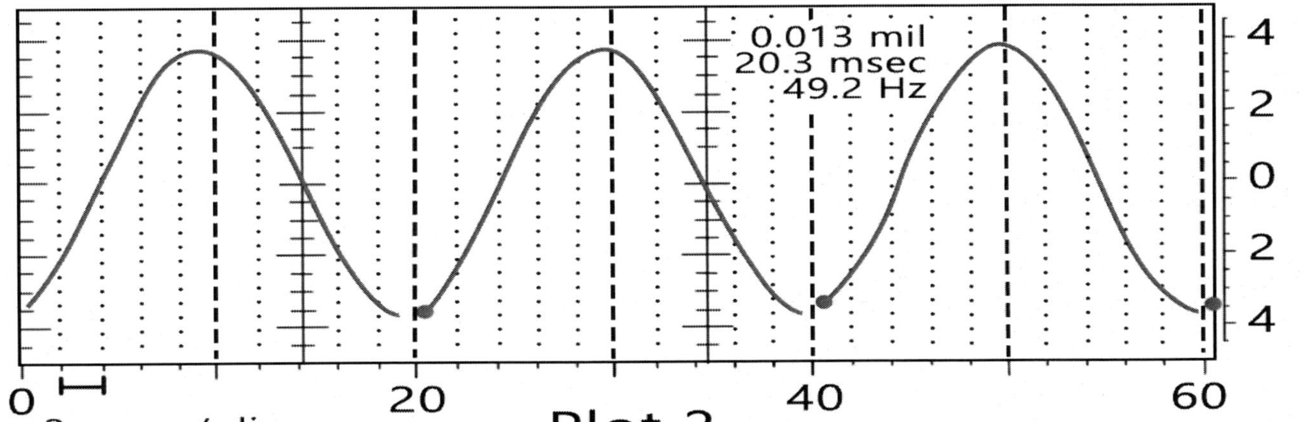
0.013 mil
20.3 msec
49.2 Hz
4
2
0
2
4
0
20
40
60
2 msec / div
Plot 2

Fault Analysis and Correction

Question 010

Study the provided plots. What is the difference between these two spectrum plots?

A. Units of amplitude and frequency

B. Calculation method of spectral amplitude

C. The lower plot has a higher F_{max}

D. All of the above

Note : ....

....

....

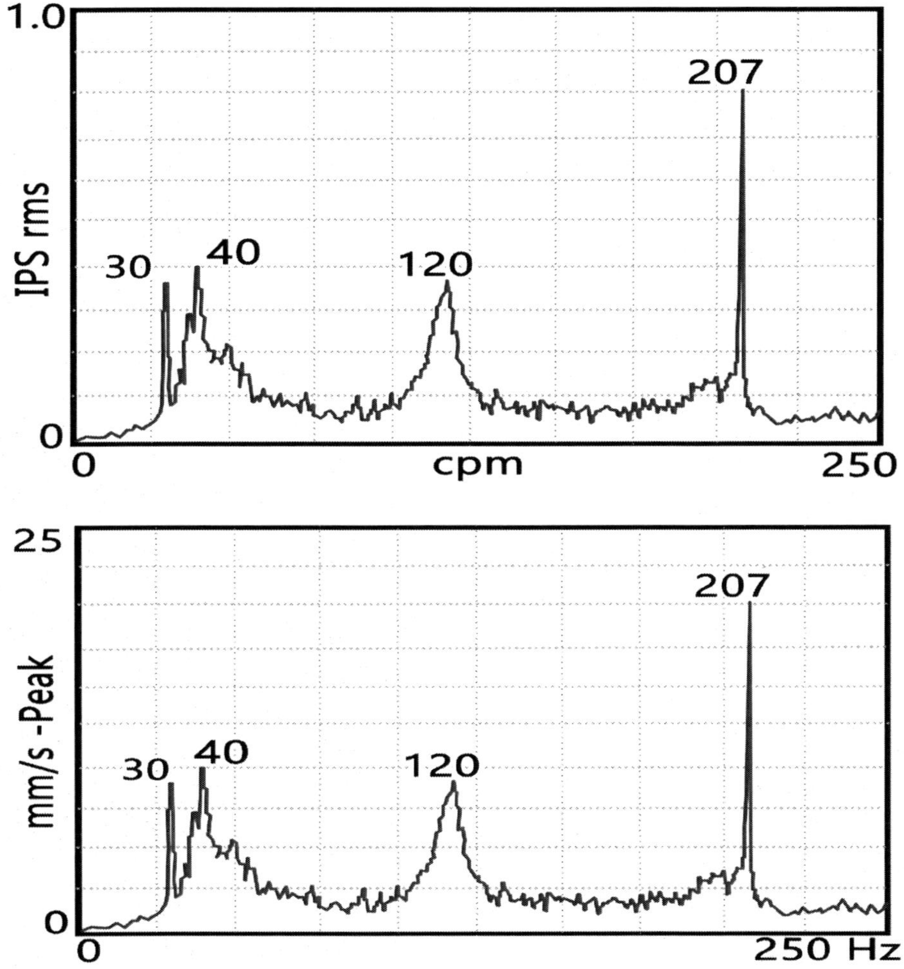
1.0
IPS rms
30
40
120
207
0
0
cpm
250
25
mm/s -Peak
30
40
120
207
0
0
250 Hz

Fault Analysis and Correction

Question 011

What is the difference between these vibration plots?

A. The type of vibration sensor.

B. Plot 2 has a higher resolution.

C. Plot 1 is before correction while Plot 2 is after correction.

D. Plot 1 is a waveform but Plot 2 is a trend.

Note : ......

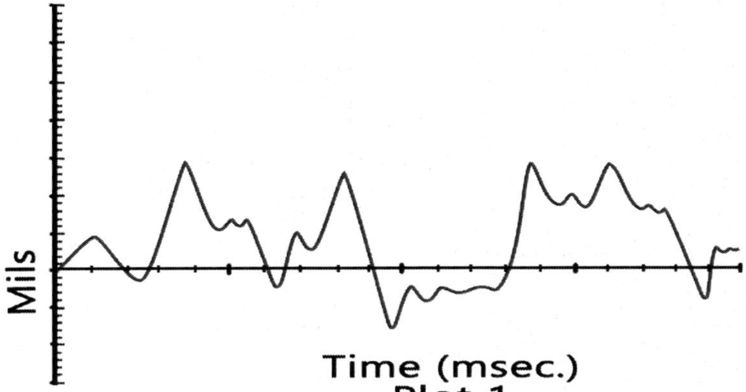

Plot 1

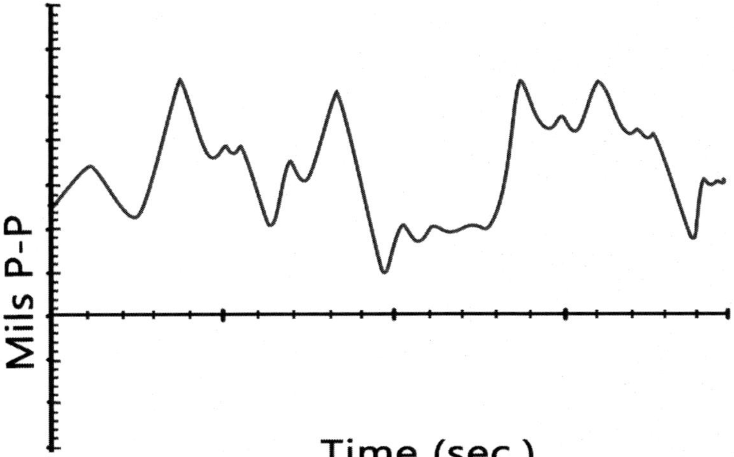

Plot 2

Fault Analysis and Correction

Question 012

Consider typical shaft vibration signals from a typical motor (e.g. 1800 rpm, 50 Hp). Which of the following is a common, correct or possible vibration frequency of a mechanical fault that you would analyze?

1. 0.005 Hz 2. -1 Hz 3. 0.05 cpm 4. 50 kHz 5. 5000 Hz

A. 1, 2 and 5

B. 2, 3 and 4

C. 4 and 5

D. None of the above

Note : ....

....

....

Fault Analysis and Correction

Question 013

Consider typical vibration signals from a typical motor (e.g. 1500 rpm, 37 kW). Which of the following is a common, correct or possible frequency of a mechanical fault that you would analyze?

1. -10 kHz **2. -5 cps p-p** **3. 0 cpm** **4. 0.5 mil p-p**

A. 1 and 2

B. 1, 2 and 3

C. 2, 3 and 4

D. None of the above

Note : ....

....

....

Fault Analysis and Correction

Question 014

What are the best descriptions for the provided waveform? Agnore normal noise and select the best from the below expressions.

A.	Harmonic	Erratic
B.	Composite	Periodic
C.	Random	Erratic
D.	Pure Sinusoidal	Non-Periodic

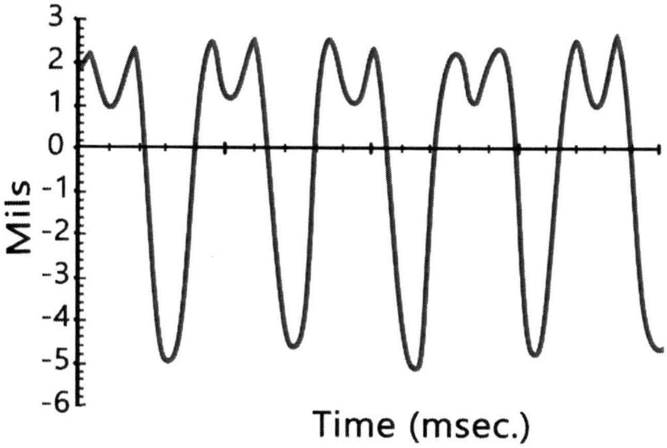

Note : ....

Fault Analysis and Correction

Question 015

What is the meaning of "integer" in the context of machinery vibration spectral analysis?

A. A whole number (with no fraction)

B. A complex number based on π

C. The integration factor (or constant) generated from a sine wave.

D. Any number between 0 and 1.

Note : ....

....

....

Fault Analysis and Correction

Question 016

What are the best descriptions for the below waveform which is taken from an induction motor? Select from the below expressions.

A.	Harmonic	Filtered
B.	Harmonic	Periodic
C.	Random	Erratic
D.	Random	Synchronous

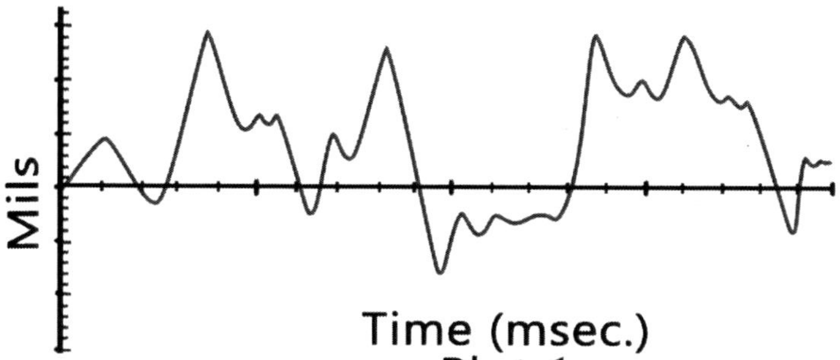

Plot 1

Note :

Fault Analysis and Correction

Question 017

What are the best descriptions for the below spectrum? Select the best from the below expressions.

A.	Ski slope	High noise
B.	Distinct components	Speed normalized
C.	Wide band noise	Side bands
D.	High flour noise	Directional

Vibration Spectrum

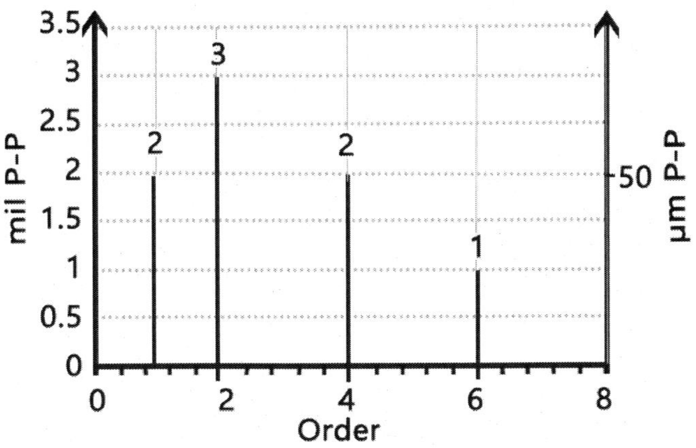

Fault Analysis and Correction

Question 018

Consider typical vibration setup for a centrifugal pump (3000 rpm, 150 kW). What is the best description for the below spectrum? Select the best from the below expressions.

A.	Ski slope	Low frequency range
B.	Distinct components	High frequency range
C.	Wide band noise	Velocity
D.	Side bands	Displacement

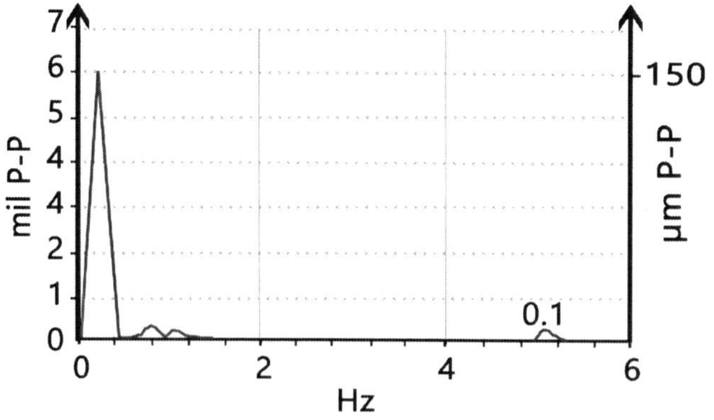

Fault Analysis and Correction

Question 019

What is the best description for the below spectrum? Select the best from the below expressions.

A. Ski slope

B. Harmonic components

C. Sidebands

D. High flour noise

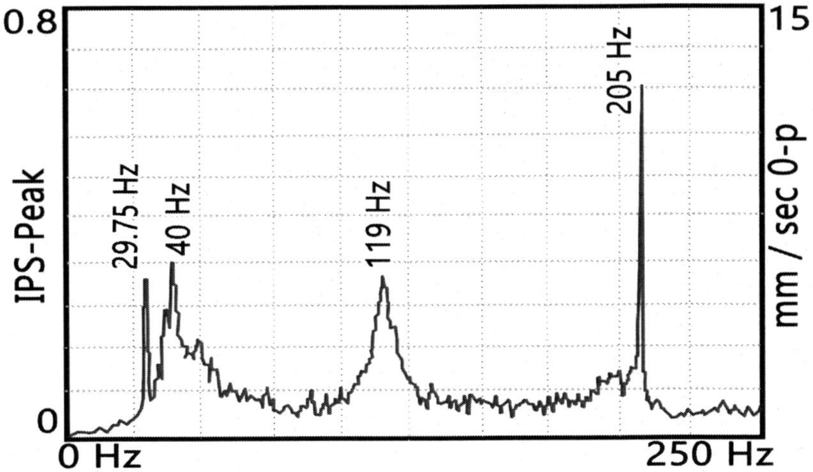

Note : ....

....

....

Fault Analysis and Correction

Question 020

A vibration specialist contractor was called to analyze high 1X vibrations on a horizontal fan at a paper factory. His recommendation was to remove the rotor for cleaning and balancing. After 11 month of operation, the fan's rotor was removed to the repair shop and was inspected. Only one side of the fan was found dirty. However, the ball bearings were found damaged. What is your assessment/ recommendation?

A. Most likely the contractor made a mistake. The defect was a bearing defect and the correction had nothing to do with rotor balancing.

B. Perform a rotor balancing check. The bearings could have failed due to the forces of mass unbalance.

C. Clean the rotor and replace the bearings. No need for balancing.

D. Couple to the motor and perform motor current analysis to identify the cause.

Fault Analysis and Correction
Question 021

These vibration signals are unfiltered time waveforms. What could be the most possible cause of each signal? Assume these plots are from similar machines and have the same data collection duration.

	1	2	3
A.	Unbalance	Background pulsation	Pump Cavitation
B.	Gear tooth defect	Shaft Misalignment	Pump Cavitation
C.	Bad sensor	Shaft Misalignment	Unbalance
D.	Bearing cage defect	Shaft bow	Electrical fault

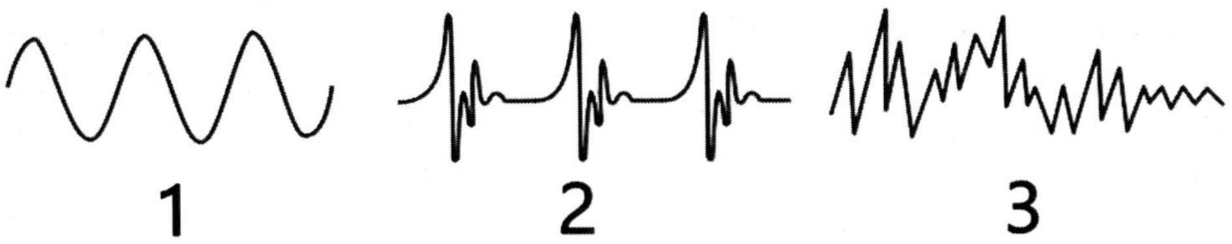

Fault Analysis and Correction

Question 022

What is a key vibration symptom that indicates a mechanical looseness problem of the parts in an induction motor?

A. Stable 2LF but unstable 4LF.

B. Clear presence of 1X and its harmonics in the spectrum.

C. Smooth AC part of the gap voltage but unstable DC part.

D. All of the above.

Note : ....

Fault Analysis and Correction
Question 023

In the field of machinery vibration diagnostics, what does each of the following acronyms stand for?

	VPF	BPF	GMF	LF
A.	Vibrating Pulse Fault	Best Performance Frequency	Newton's Second Law of Motion	Limited Fault
B.	Vane Pass Frequency	Blade Pass Frequency	Gear Mesh Frequency	Line Frequency
C.	Velocity Pass Frequency	Bearing Pass Frequency	Gravity Motion Frequency	Logarithmic Force
D.	Vapor Pressure Fault	Boiling Pressure Fault	Gama Motion Fault	Linear Feedback

Fault Analysis and Correction

Question 024

Assume a typical horizontal motor driving a water pump. What are typical vibration frequencies generated by the following faults? "X" refers to shaft running speed.

	Motor Rotor Mass Unbalance	Shaft Misalignment (at the coupling)	Pump Bent Shaft	Pump Rotor Mass Unbalance
A.	(1X)	(1X, 2X, 3X)	(1X)	(1X)
B.	(2X)	(0.1X, 0.2X, 0.3X)	(1/X)	(1X)
C.	(1X, 1/2X, 1/3X)	(1X, 2X, 3X)	(1X, 2X, 3X)	(1X, 1/2X, 1/3X)
D.	(1X, 2X, 3X)	(1X)	(1X)	(1X, 2X, 3X)

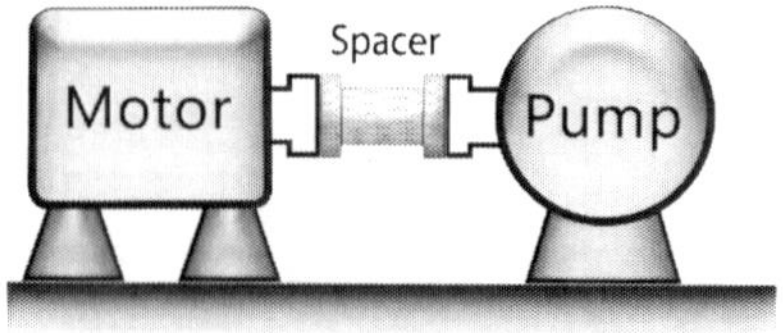

Fault Analysis and Correction

Question 025

What are the typical vibration frequencies generated by the following faults?
Assume typical industrial machines. Select the best option from the below choices.

	Pump Hydraulic Problems	Motor Electromagnetic Faults
A.	1/X	1X and/or LF
B.	VPF	1X, 2X and/or 2LF
C.	GMF	60 Hz and 120 Hz
D.	5X	120 Hz ±GMF

Fault Analysis and Correction

Question 026

Consider a centrifugal pump. Which of the following can be a corrective action for rotor unbalance, shaft misalignment and cavitation?

	Rotor Unbalance	Shaft Misalignment	Cavitation
A.	Replace bearing	Improve support	Lubricate the impeller
B.	Align motor to pump	Replace bearing	Decrease the inlet pressure
C.	Tighten hold down bolts	Add or remove weight	Balance the impeller
D.	Add or remove weight	Align shafts	Provide design pressure at suction side

Fault Analysis and Correction

Question 027

Consider a 20 Hp motor driving a horizontal pump. Which of the following can be a corrective action for resonance, bearing damage and looseness?

	Resonance	Bearing Damage	Looseness
A.	Increase resolution	Align coupling hubs	Add or remove weight
B.	Improve supports	Replace bearing	Tighten bolts
C.	Align shafts	Tighten bolts	Increase resolution
D.	Increase F_{max}	Drain the lube oil	Increase suction pressure

Note : ....

Fault Analysis and Correction

Question 028

A motor was reported with high vibrations. After inspection, the millwright stated that the motor's shaft is too high compared to the driven equipment. What do you recommed as a corrective action in this case?

A. Align the shafts.

B. Balance the shafts.

C. Replace the shafts.

D. Lower the resolution.

Note : ....

....

....

Fault Analysis and Correction
Question 029

For the next plot, what is the closest approximation of amplitude and frequency?

	Amplitude	Frequency
A.	0.8 0-p Vol	7350 cpm
B.	0.7 p-p Vol	8.17 msec
C.	1.5 0-p Vol	120 Hz
D.	6.125 p-p Vol	7200 cpm

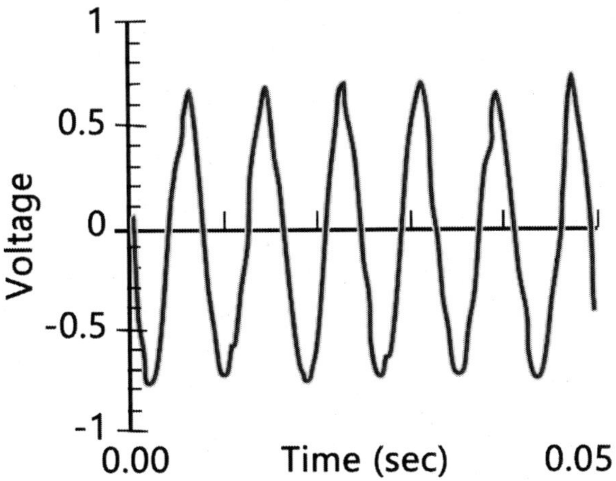

Note : ....

....

....

Fault Analysis and Correction

Question 030

For the next plot, consider the dominant signal. What is the best approximation for the period? What is phase angle considering the starting time as a reference? Hint: For phase, find what part of a cycle is between the reference and the peak.

	Period (msec)	Phase (deg)
A.	122	100
B.	0.008	0
C.	8.16	280
D.	8.16	0.05

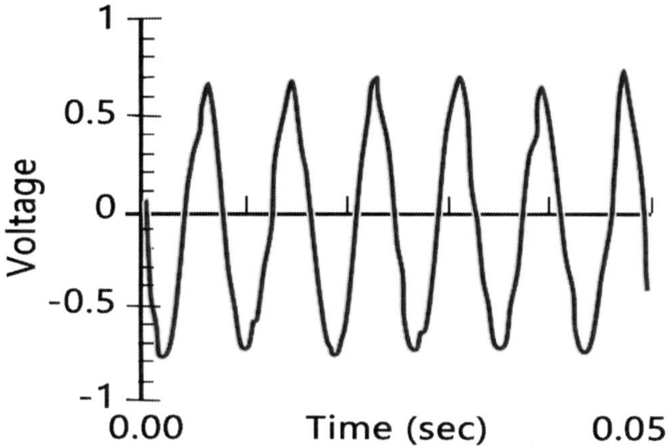

Note :

Fault Analysis and Correction

Question 031

Study the next spectrum. What are the frequencies of the two highest amplitudes? Shaft speed is 1780 rpm.

A. 1X and 2X

B. 8X and 9X

C. 1X and 8X

D. 1780 cpm and 16020 cpm

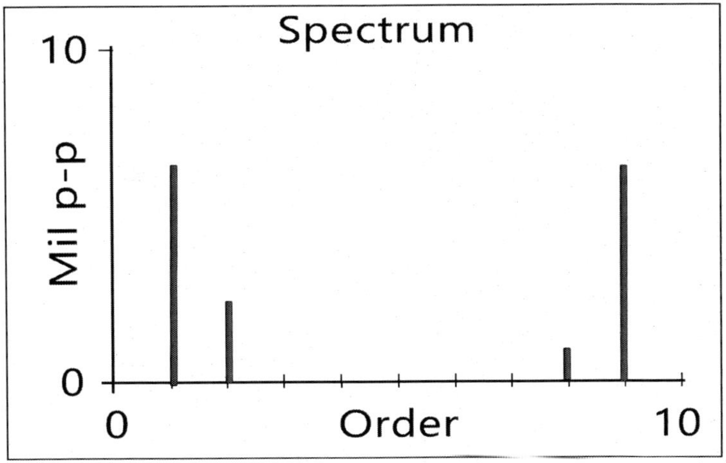

Note : ....

....

....

Fault Analysis and Correction

Question 032

Study the vibration spectrum from Cooling Water Pump A. The shaft speed is 6000 rpm or 100 Hz. What is the vibration amplitude of 0.5X and 1X?

	0.5X	1X
A.	2.4 mil p-p	1.2 mil p-p
B.	0.08 mil p-p	1.1 mil p-p
C.	0.8 mil p-p	2.4 mil p-p
D.	1.1 mil p-p	2.4 mil p-p

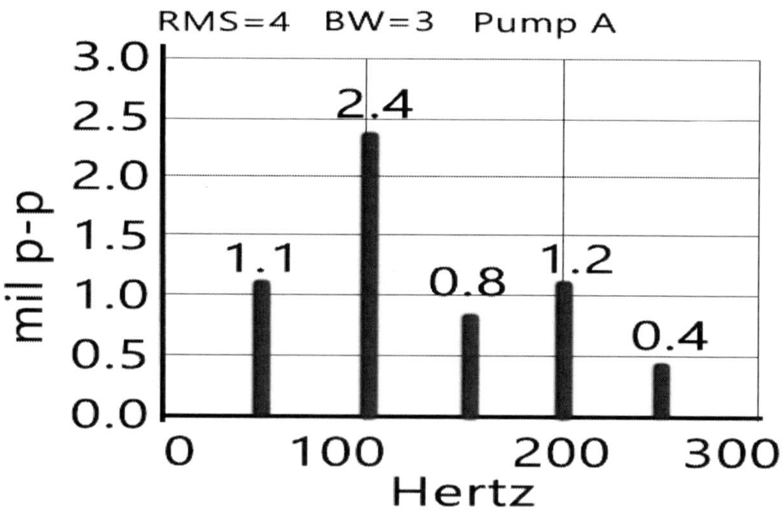

Fault Analysis and Correction
Question 033

For the next spectrum, what are the frequencies of the two components? Shaft speed is 3580 rpm. Hint: convert the units to cpm.

A. 7.5X and 1X

B. 1X and 2X

C. 1X and 8X

D. 1800 cpm and 14400 cpm

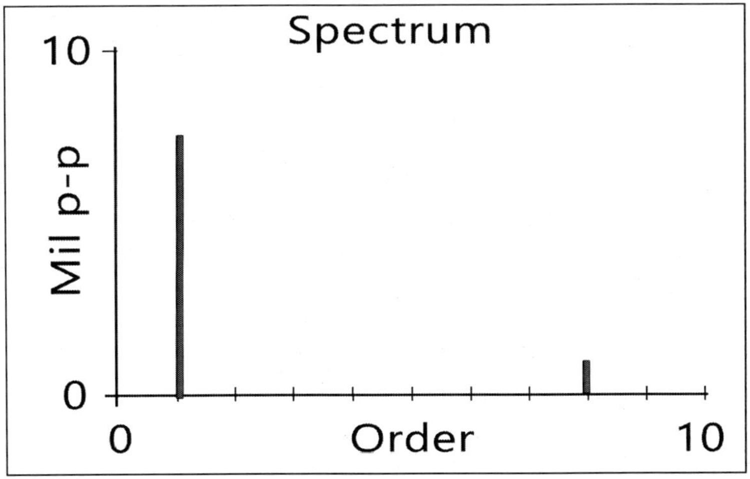

Note : ....

Fault Analysis and Correction

Question 034

For the next spectrum, what are the frequencies of the two tallest spectral components? Shaft speed is 3580 rpm. Hint: You may need to convert the units to cpm.

A. 1X and 32220 cpm

B. 1800 cpm and 14400 cpm

C. 3580 cpm and 10X

D. 3580 cpm and 36000 cpm

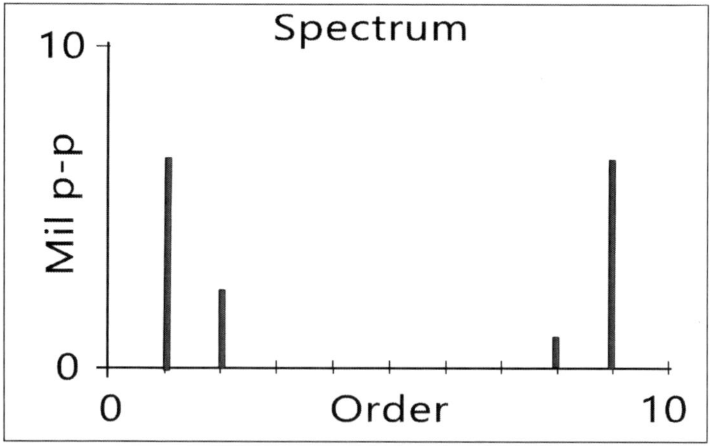

Note :

Fault Analysis and Correction

Question 035

Study the next spectrum. What are the frequencies of the two peaks at the high frequency end of the spectrum? Shaft speed is 1780 rpm. Hint: You may need to convert the units to cpm and Hz.

A. 1X and 9X

B. 14240 cpm and 267 Hz

C. 2X and 8X

D. 297 Hz and 267 Hz

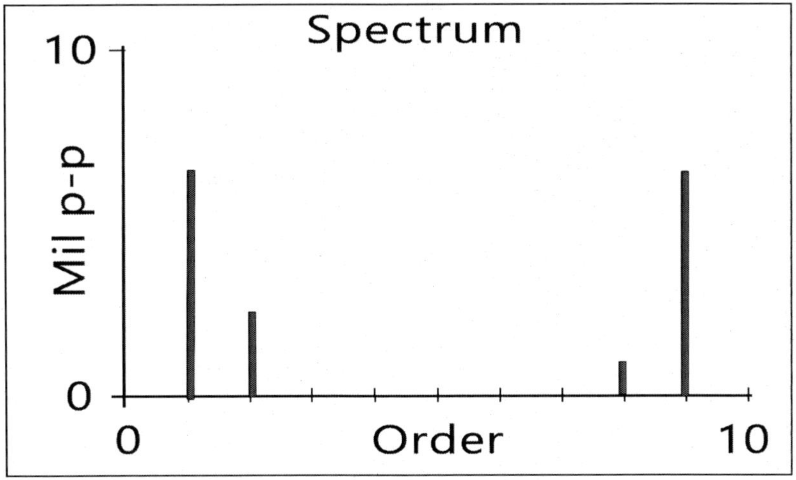

Note : ....

....

....

Fault Analysis and Correction

Question 036

Study the next vibration spectrum. What is the frequency and amplitude for the key component? The original data covers 800 lines over 800 Hz. However, only data up to 400 Hz is shown in this spectrum. No major peak is found after 400 Hz.

A. 1.5 Hz and 0.100 ips rms

B. 120 Hz and 0.095 ips rms

C. 400 Hz and 0.100 ips rms

D. 30 Hz and 0.100 mm/sec rms

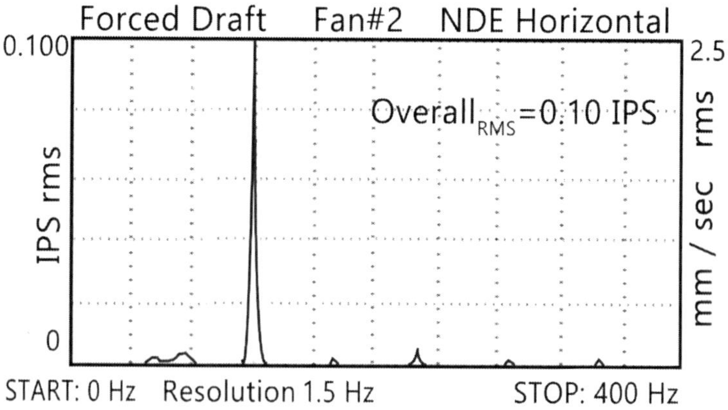

Note :

Fault Analysis and Correction

Question 037

Study the next vibration frequency spectrum measured from a shaft of a large horizontal pump running at 6300 rpm. What is the period of the first vibration component? What is the order of the largest amplitude component? Hint: p = 1 / f.

	Period	Order
A.	19 msec	1 X
B.	0.20 sec	2.4 X
C.	2.4 msec	105 X
D.	1.1 msec	10 X

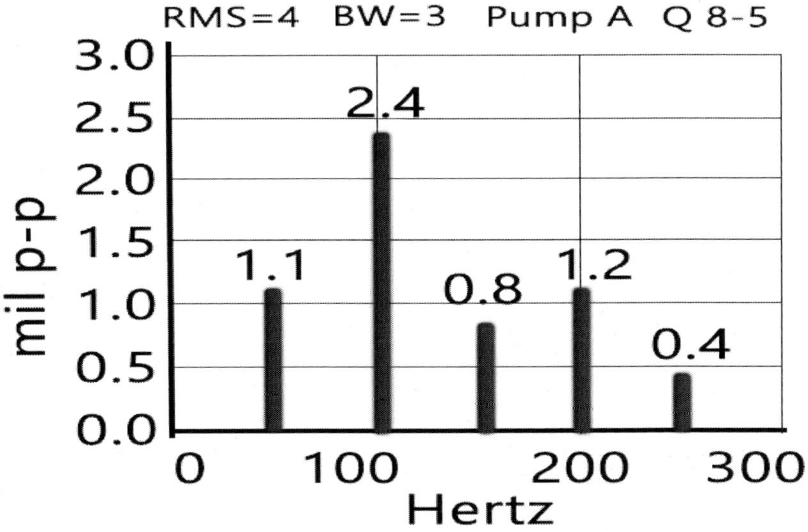

Fault Analysis and Correction

Question 038

Study the next vibration frequency spectrum measured from a bearing housing. What is the frequency of the tallest vibration component? What is its order? Expect some rounding.

	Frequency	Order
A.	107 Hz	Insufficient information is given.
B.	107 Hz	1X
C.	6420 cpm	2X
D.	2.24 ips	2.1 X

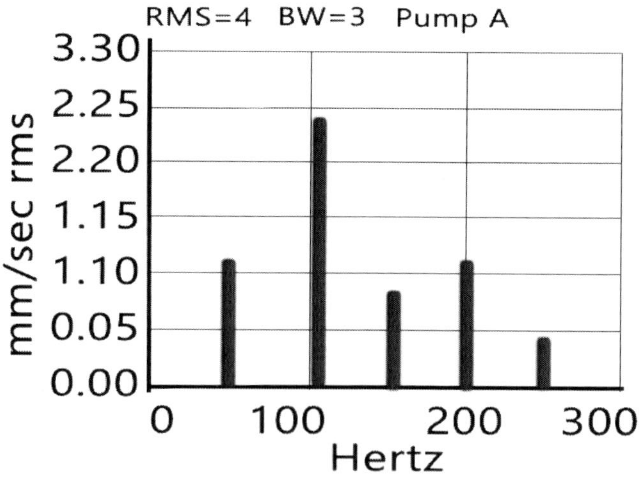

Fault Analysis and Correction
Question 039

Study the next vibration frequency spectrum measured from a shaft of a large horizontal pump. What is the period of the tallest vibration component? What is its phase? What is the cause of this vibration component? Expect some rounding.

	Period	Phase	Cause
A.	107 sec	2.4°	Unbalance
B.	9.5 msec	Insufficient information	Insufficient information
C.	107 Hz	90°	Misalignment
D.	Insufficient information	180°	Bearing Damage

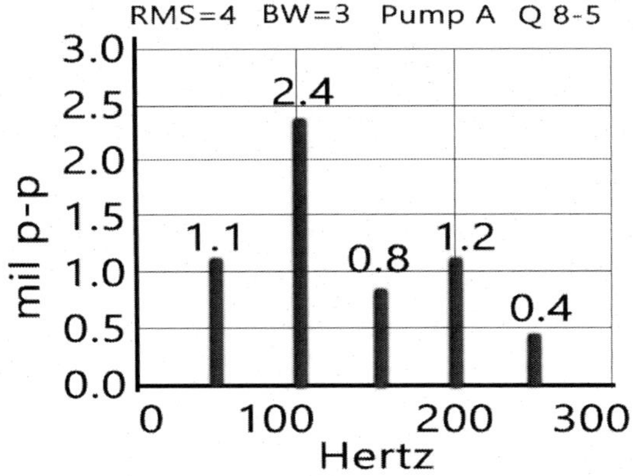

Fault Analysis and Correction

Question 040

Consider a horizontal centrifugal fan. What is the strongest vibration symptom of a typical shaft misalignment problem? Select from the given options.

A. In waveforms, there is a sharp spike component.

B. In frequency spectra, there are major components at 1 and 2 times running speed.

C. In frequency spectra, as the speed increases, window factor increases.

D. In frequency spectra, there is a major component at 60 Hz.

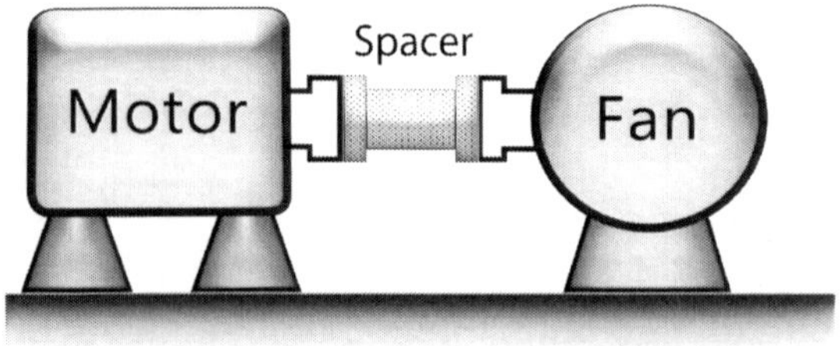

Fault Analysis and Correction

Question 041

Study the next vibration spectrum. Assume Window Factor =1.5 and Resolution Lines are 400. This data is for a DC motor running at full speed of 3599 rpm. What is the method of amplitude calculation and phase angle of the highest peak?

	Method	Phase Angle
A.	Hz/div	20 Deg
B.	Peak	Unknown
C.	RMS	60 Deg
D.	Inch/sec	0.4 ips

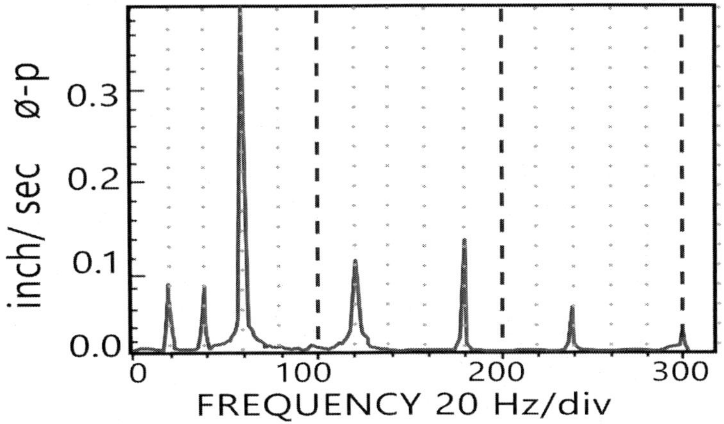

Note :

Fault Analysis and Correction

Question 042

For the next spectrum, what are the 2X and 3X amplitudes? The machine is running at 60 Hz but under 50% load. Expect some rounding.

	2X (inch/sec rms)	3X (inch/sec rms)
A.	0.08	0.03
B.	0.18	0.09
C.	0.36	0.18
D.	60	120

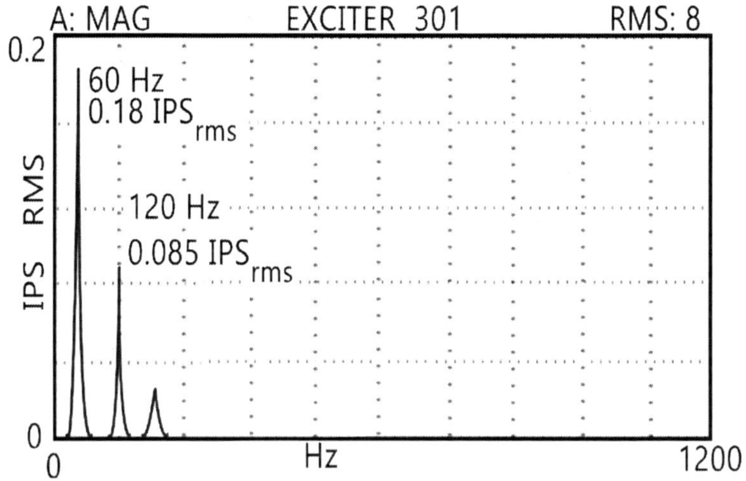

Note :

Fault Analysis and Correction

Question 043

For the next spectrum, what are the 2X and 4X frequencies? Assume the machine is running at 60 Hz but under a 50% load. Expect some rounding.

	2X	4X
A.	30 Hz	60 Hz
B.	3600 cpm	7200 cpm
C.	60 Hz	1200 Hz
D.	None of the above	

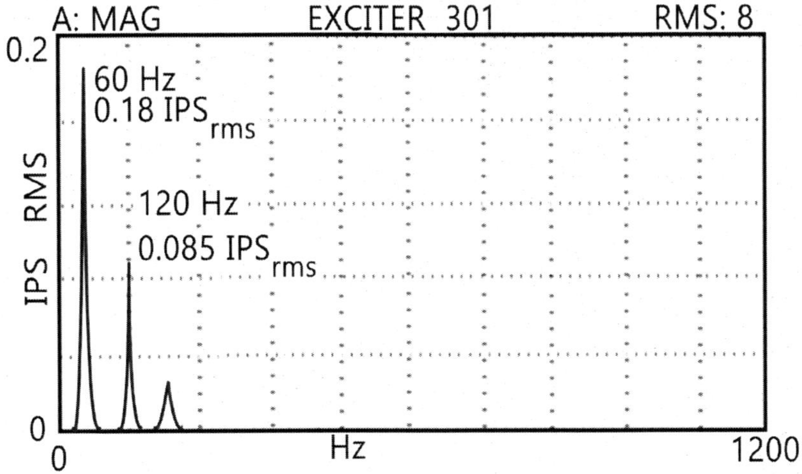

Note :

Fault Analysis and Correction

Question 044

For the next vibration spectrum, what is the 2X frequency and amplitude? Assume the machine is running at 7199 rpm but under 105% load. Expect some number rounding.

	Frequency (Hz)	Amplitude (ips rms)
A.	60	0.18
B.	180	0.3
C.	120	0.09
D.	240	0.00

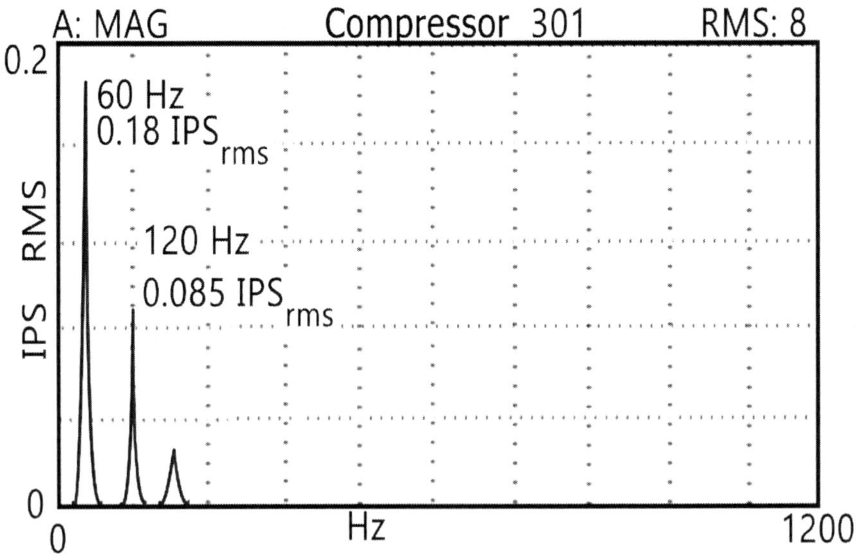

Fault Analysis and Correction
Question 045

For the next vibration spectrum, what are the 1X and 2X frequencies? The compressor is running at 3600 rpm but under 120% of the design flow.

	1X (Hz)	2X (Hz)
A.	3600	7200
B.	60	120
C.	30	15
D.	0.18	0.084

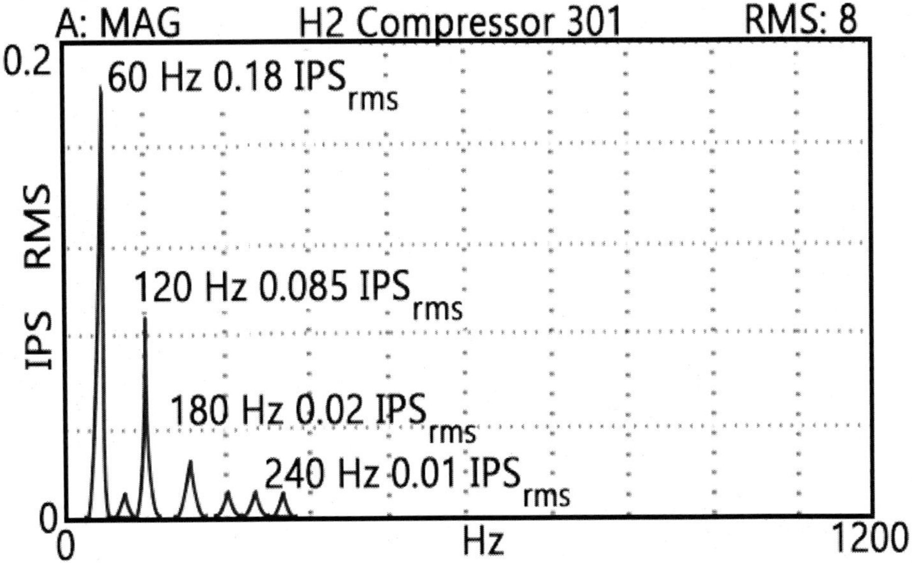

Fault Analysis and Correction

Question 046

For the next vibration spectrum, what are the 2X and 4X amplitudes? The compressor is running at 3600 rpm but under 50% of the design pressure. Expect some number rounding.

	2X (inch/sec rms)	4X (inch/sec rms)
A.	0.09	0.01
B.	0.18	0.08
C.	0.36	0.16
D.	3600	7200

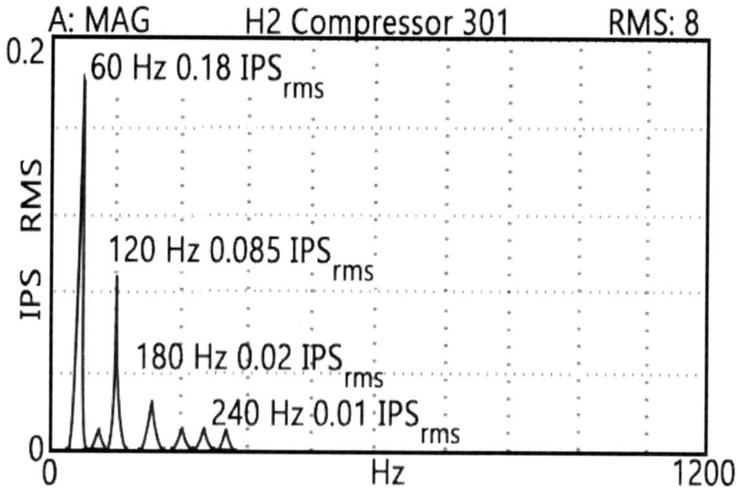

Note : ....

Fault Analysis and Correction

Question 047

For the next vibration plot, what are the 1/2X and 2X amplitudes? The pump is running at 7200 rpm but its bearings are at a 50% clearance more than the acceptable value. Expect some number rounding. Hint: convert speed to Hz then locate the 1X camponent.

	1/2X (ips 0-p)	2X (ips 0-p)
A.	0.08	0.08
B.	0.01	0.39
C.	0.12	0.06
D.	0.39	0.07

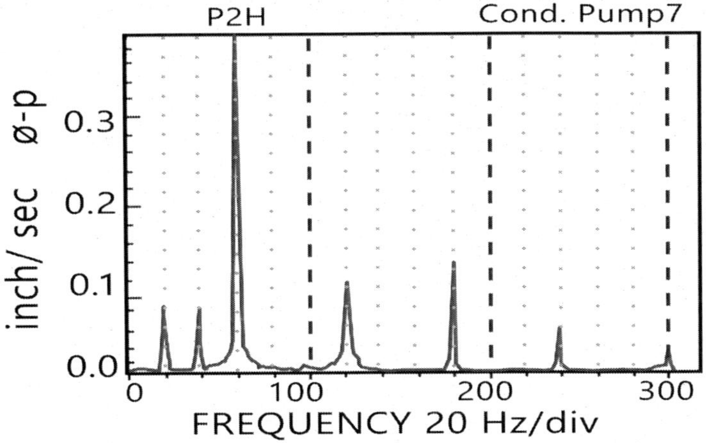

Note :

Fault Analysis and Correction

Question 048

For the next vibration plot, what are the frequencies indicated by P#3 and P#4? The pump is running at 7200 rpm and data collection time is 0.8 seconds. Hint: convert the speed to Hz then locate 1X component.

	P#3	P#4
A.	1X	2X
B.	0.5X	1X
C.	1X	1.5X
D.	4X	5X

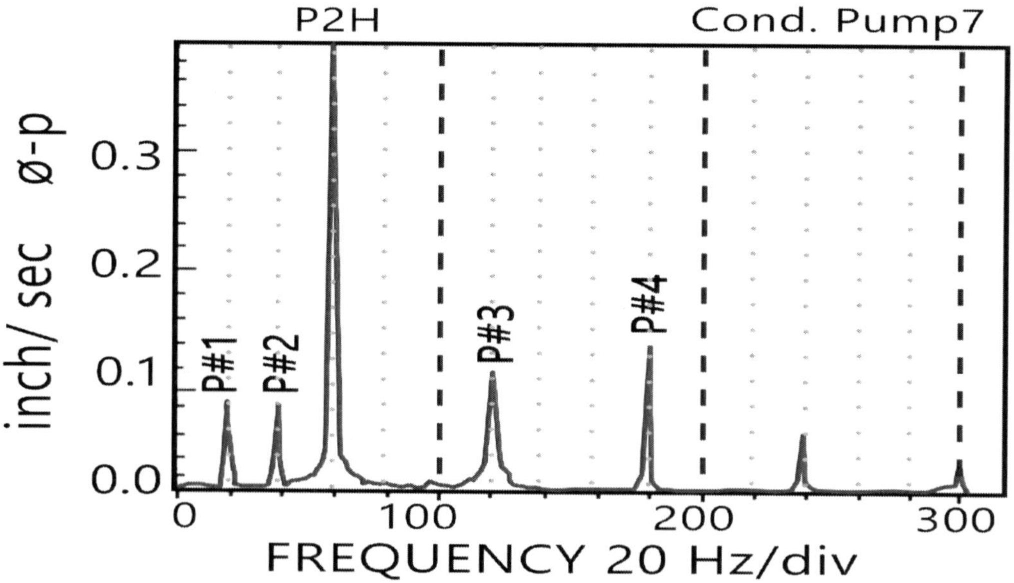

Fault Analysis and Correction

Question 049

For the next vibration plot, what are the frequencies indicated by P#1 and P#4? The pump is running at 3598 rpm and the window factor = 1.5.

	P#1	P#4
A.	1X	9X
B.	0.1	0.14X
C.	1/3 X	3X
D.	20X	180X

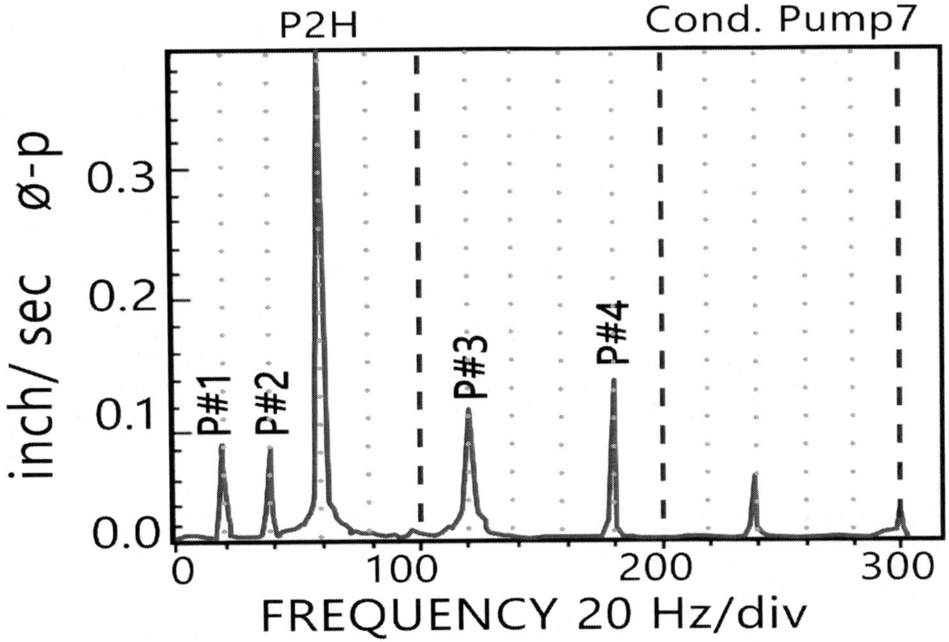

Fault Analysis and Correction

Question 050

Study the vibration waveform from a pump. The shaft speed is 998 rpm. What is the dominant vibration frequency? Expect some approximation. Hint: find the frequency from the plot in Hz then convert it to cpm. To find order use the given shaft speed.

A. 5.1 X B. 21.25 cpm C. 107.5 cpm D. 6450 cpm

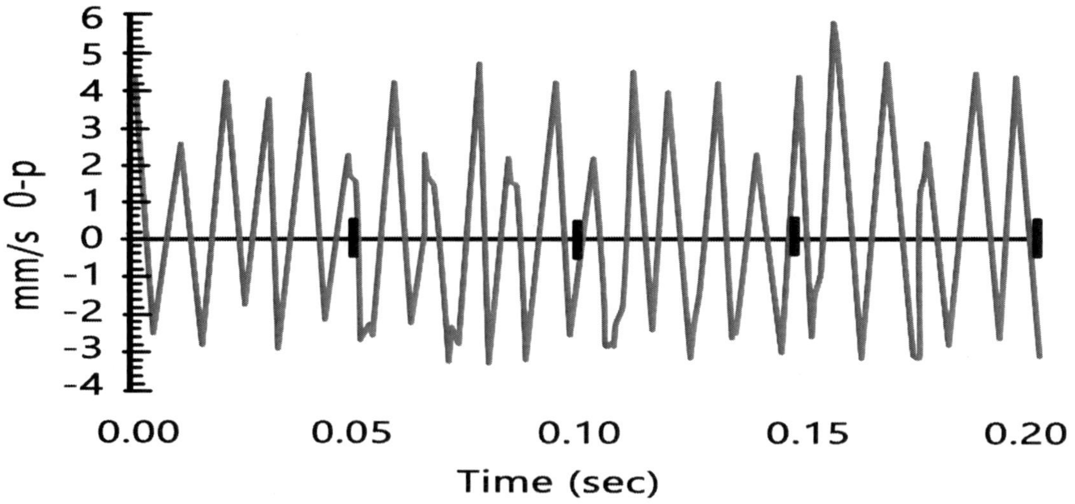

Fault Analysis and Correction

Question 051

Study the vibration waveform which was collected from a pump. The shaft speed is 1780 rpm. What is the dominant vibration frequency order? Hint: find the frequency from the plot in Hz then convert it to cpm. To find order use the given shaft speed.

A. 5.1 X   B. 21.25 X   C. 3.62 X   D. 107.5 cpm

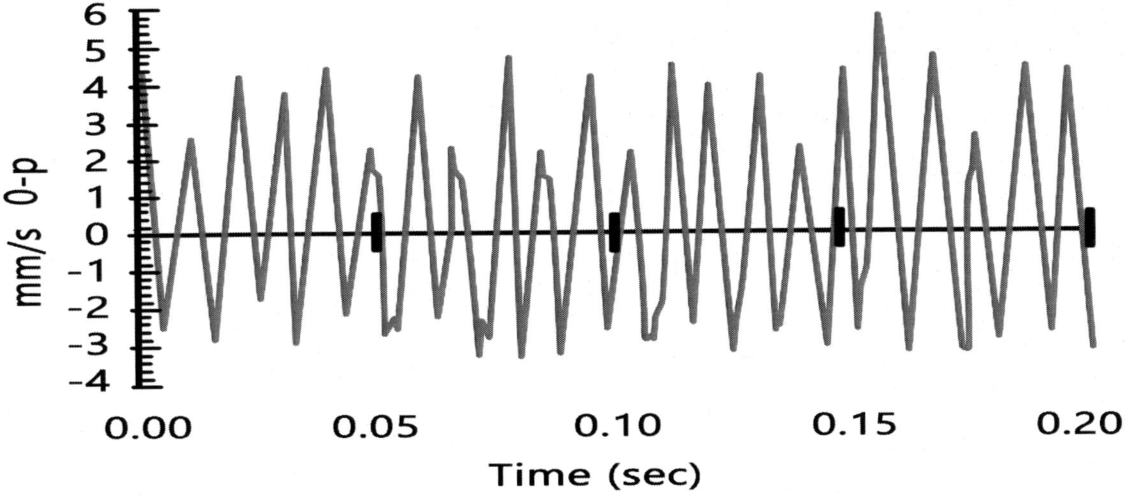

Fault Analysis and Correction

Question 052

Study the vibration spectrum from Boiler Feed Water Pump A. The shaft speed is 6000 rpm or 100 Hz. What is the vibration amplitude of 1X and 2X?

	1X	2X
A.	1.1 mil p-p	2.4 mil p-p
B.	2.4 mil 0-p	0.8 mil 0-p
C.	2.4 mil p-p	0.7 mil p-p
D.	2.4 mil p-p	1.2 mil p-p

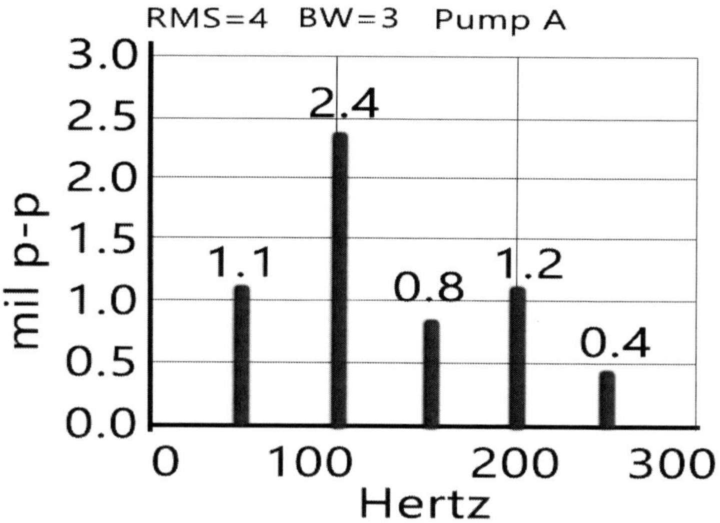

Fault Analysis and Correction

Question 053

Which waveform is the most probable to indicate a severely defective ball bearing?

A. 1 B. 2 C. 3 D. 4

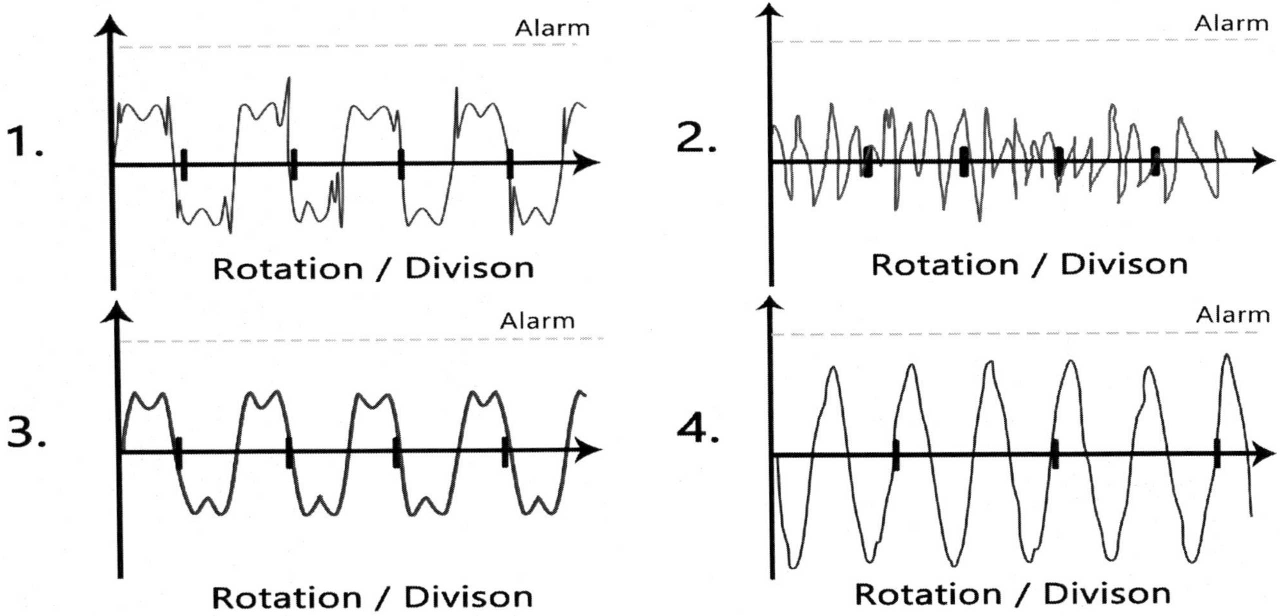

Fault Analysis and Correction

Question 054

Which waveform is the most probable to indicate a possible shaft misalignment?

A. 1

B. 2

C. 3

D. 4

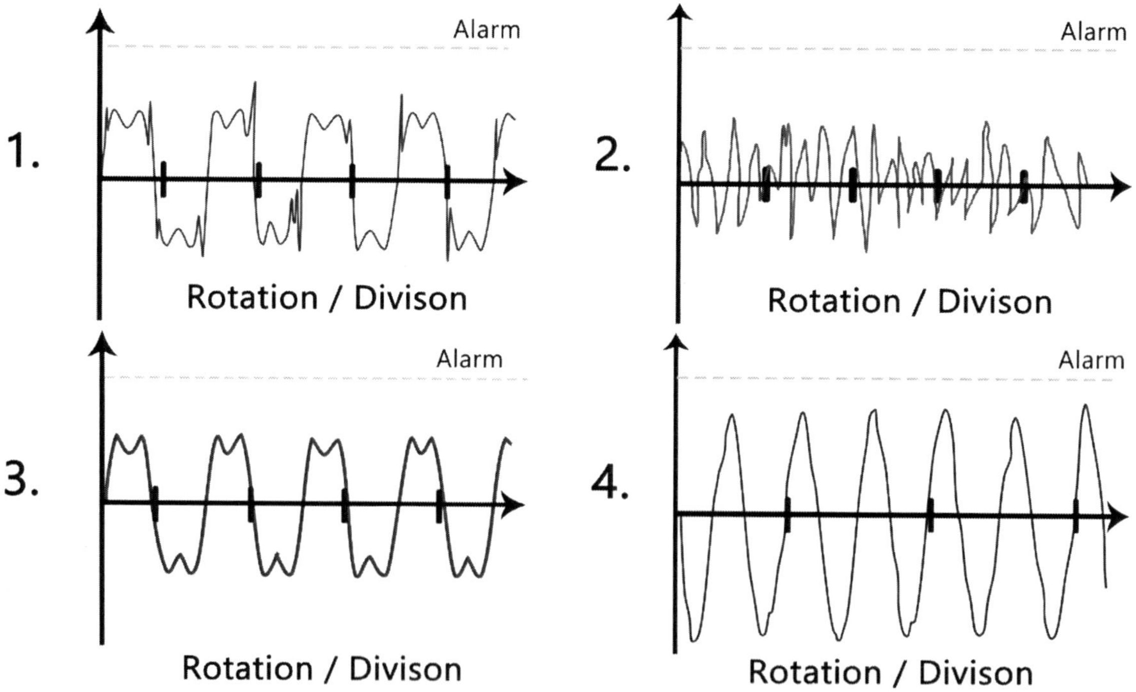

Fault Analysis and Correction

Question 055

Which waveform possibly indicates a severe hydraulic turbulence problem of a centrifugal pump?

A. 1

B. 2

C. 3

D. 4

1.

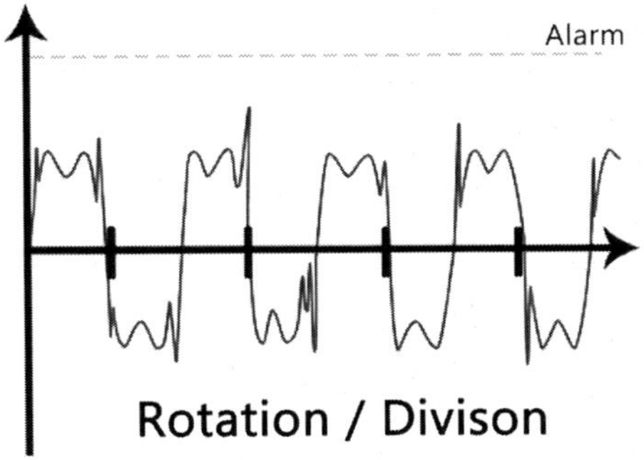

2.

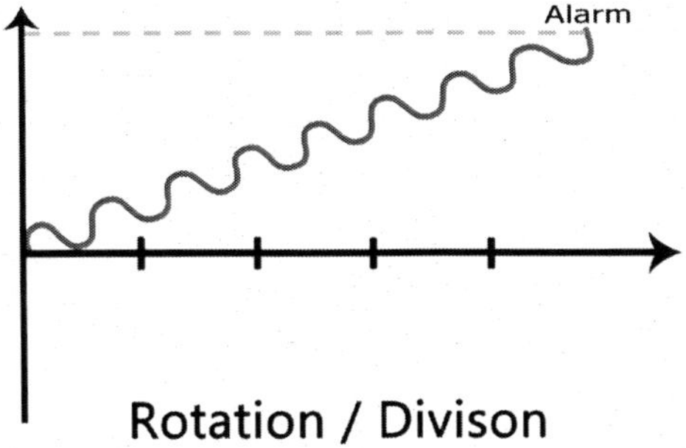

3.

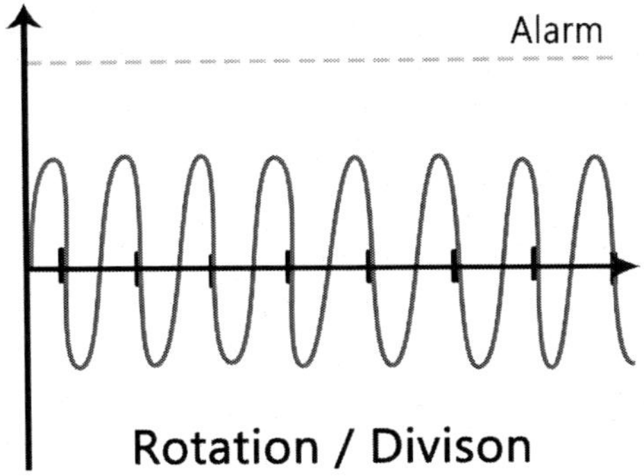

4.

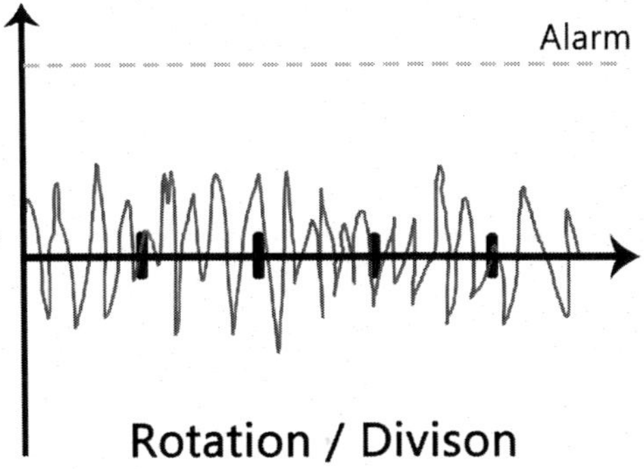

Fault Analysis and Correction

Question 056

Which waveform is the most probable indication for a rotor unbalance for a fan running at 60 Hz?

A. 1

B. 2

C. 3

D. 4

1.

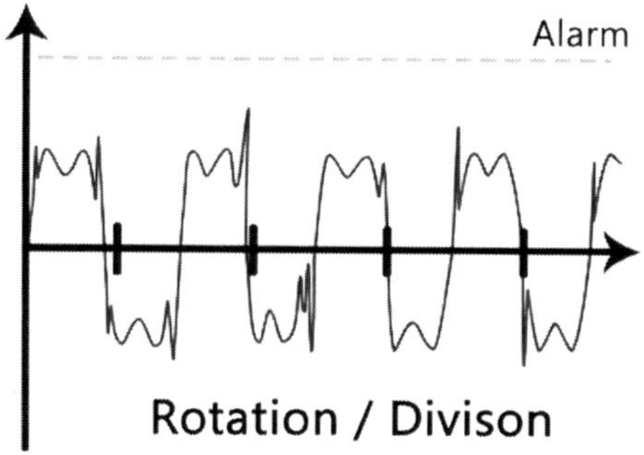

2.

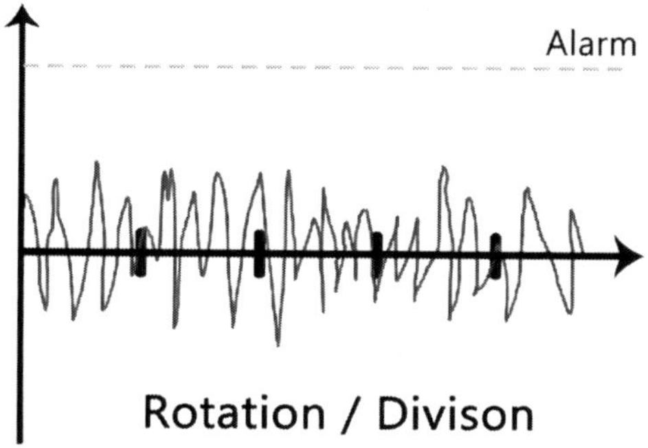

3.

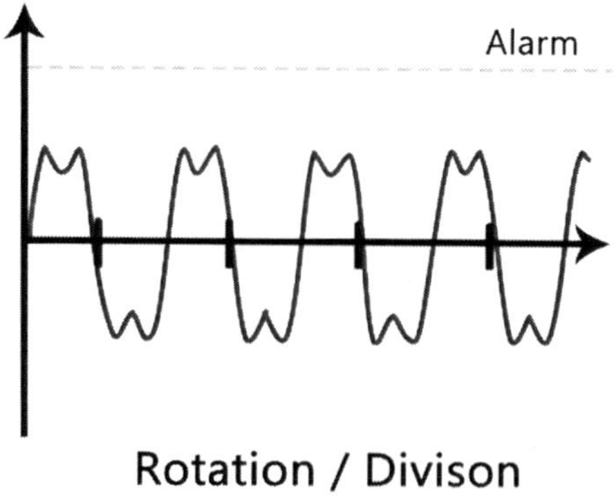

4.

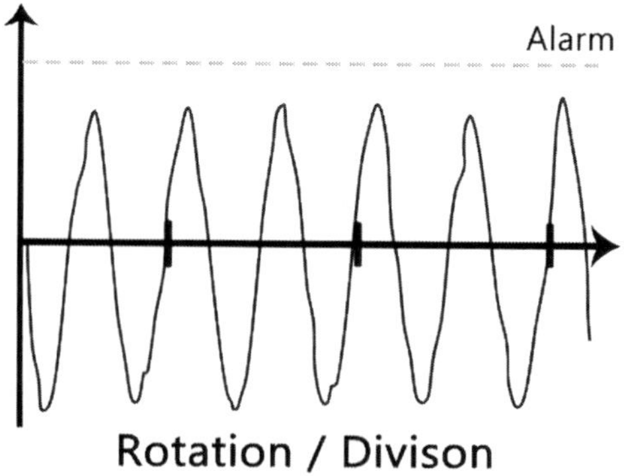

Fault Analysis and Correction

Question 057

Which spectrum possibly indicates a defective ball bearing?

A. 1

B. 2

C. 3

D. 4

1.

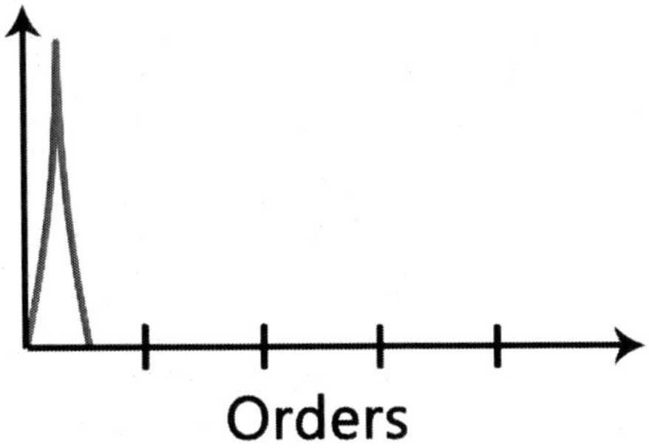

2.

3.

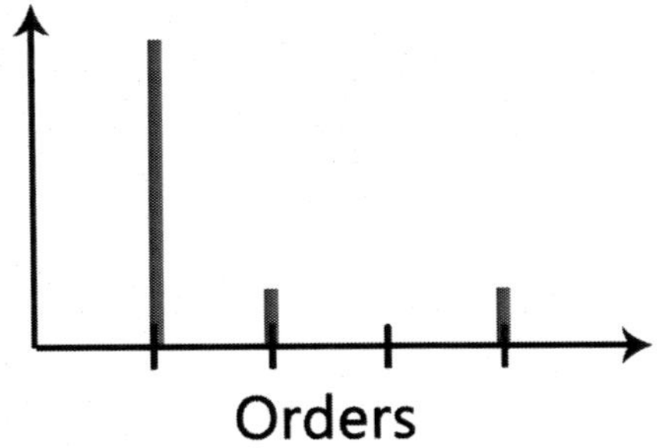

4.

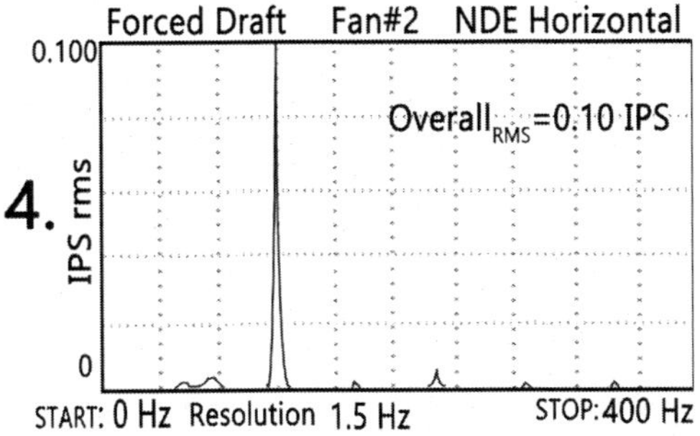

Fault Analysis and Correction

Question 058

Which vibration plot is the most probable indication for data collection problem?

A. 1

B. 2

C. 3

D. 4

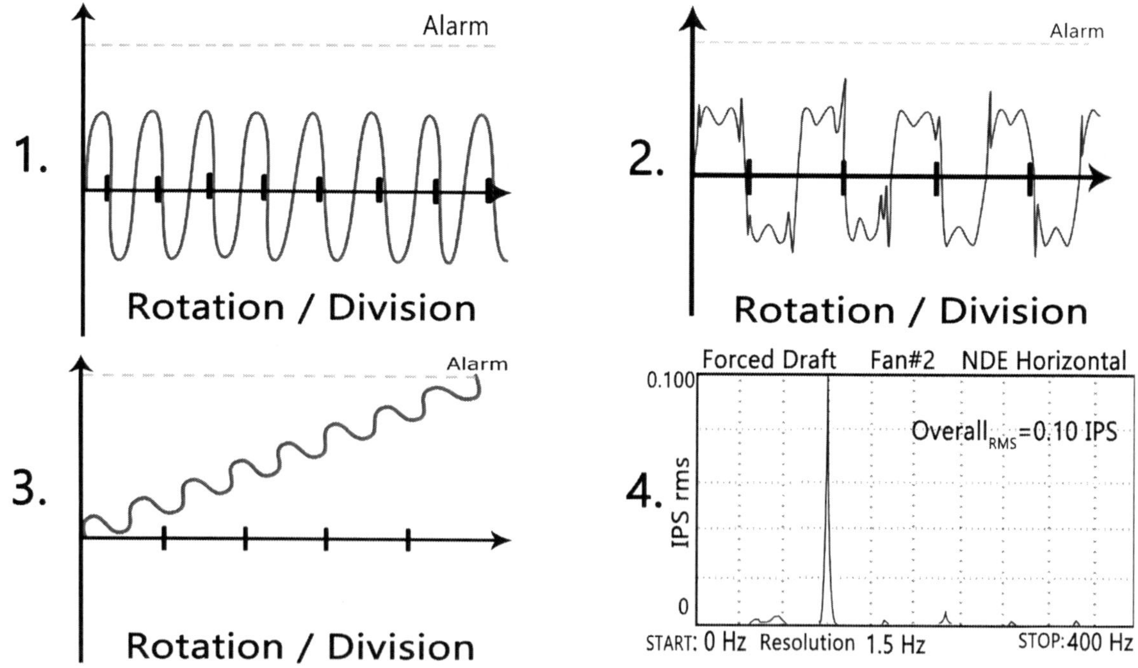

Fault Analysis and Correction
Question 059

Which waveform is the most probable indication for a rotor unbalance for a fan running at 60 Hz?

A. 1

B. 2

C. 3

D. 4

1.
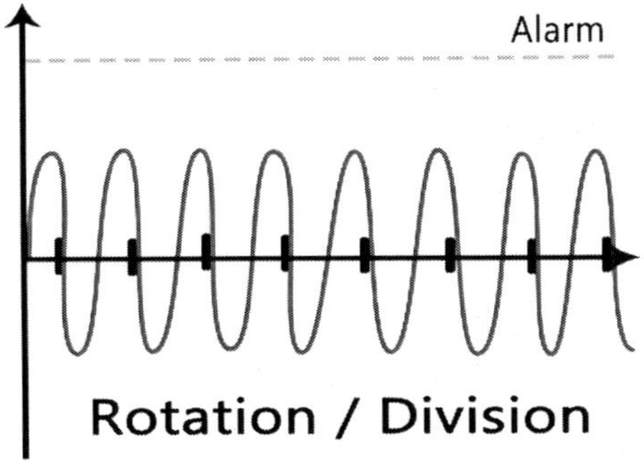

2.
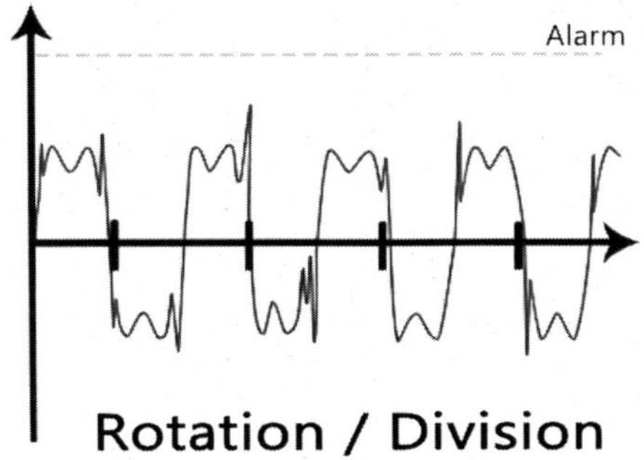

3.
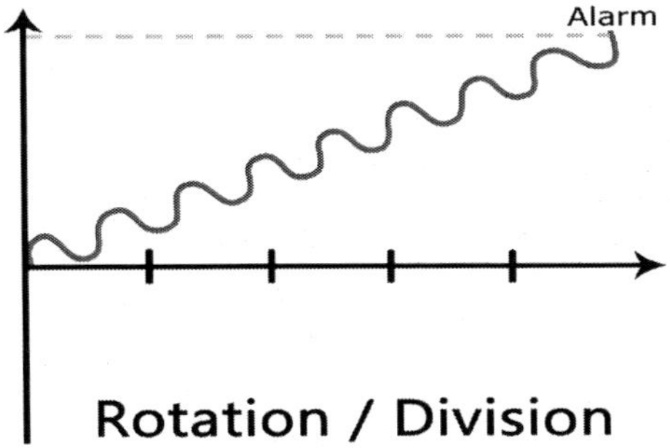

4.
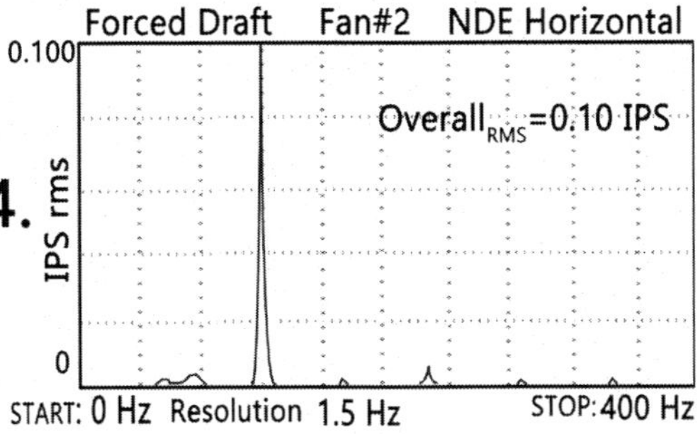

Fault Analysis and Correction

Question 060

What is a common meaning of "phase" referred to during the vibration analysis of electric motors?

A. The number of electric cables used to provide power to the motor. Some motors are single phase while others are 3 phase.

B. The time difference between a vibration peak and a reference signal. The time is usually converted to an angle out of 360 degrees.

C. Any stage in the life cycle of an electric motor. A motor can be at a design phase or life end phase.

D. A and B.

Note : ....

....

....

Fault Analysis and Correction

Question 061

Which vibration plot type is the most associated with phase angle calculations? Select the best from the given options.

A. Frequency spectrum

B. Overall amplitude trend

C. Filtered waveform

D. Unfiltered waveform

Note : ....

Fault Analysis and Correction

Question 062

What is the most commonly used unit for phase in the field of machinery vibration analysis?

A. msec

B. Degree

C. Turn

D. mVolt

Note : ....

....

....

Fault Analysis and Correction

Question 063

Which axis and plot are associated with the calculation of the phase angle in the field of machinery vibration?

A. Frequency axis in a normal frequency spectrum

B. Time axis in a filtered waveform plot

C. Axial position axis in a waveform measured from Non-Drive End

D. Y-axis of motor starting current plot.

Note : ....

Fault Analysis and Correction

Question 064

Which of the following is NOT a true statement about phase when you analyze machinery vibration?

A. Phase is used in performing rotor balancing on the site or at repair shops.

B. Phase can be a time difference between the peaks of two vibration signals in the horizontal and vertical measurements.

C. Phase angles can be calculated for shaft vibration and case vibration signals.

D. Overall phase angle is calculated from an overall (unfiltered) waveform. Phase for 1X is calculated from a 1X filtered waveform.

Note : ....

Fault Analysis and Correction

Question 065

Assume a vibration waveform signal is collected from rotating machine. It is filtered to a frequency matching with the shaft speed. How to find the phase for this signal?

A. By calculating the time difference between the reference signal and the first next positive peak.

B. By calculating the period of 1X then multiplying it by 60.

C. By calculating the period of 1X then dividing it by 60.

D. By calculating the time between the positive peak of 1 X and the next peak.

Note : ....

....

....

Fault Analysis and Correction

Question 066

Study the next vibration signal which is collected from a two pole motor. How many vibration cycles are captured in this waveform?

A. 50 B. 3 C. 4 D. 49.2

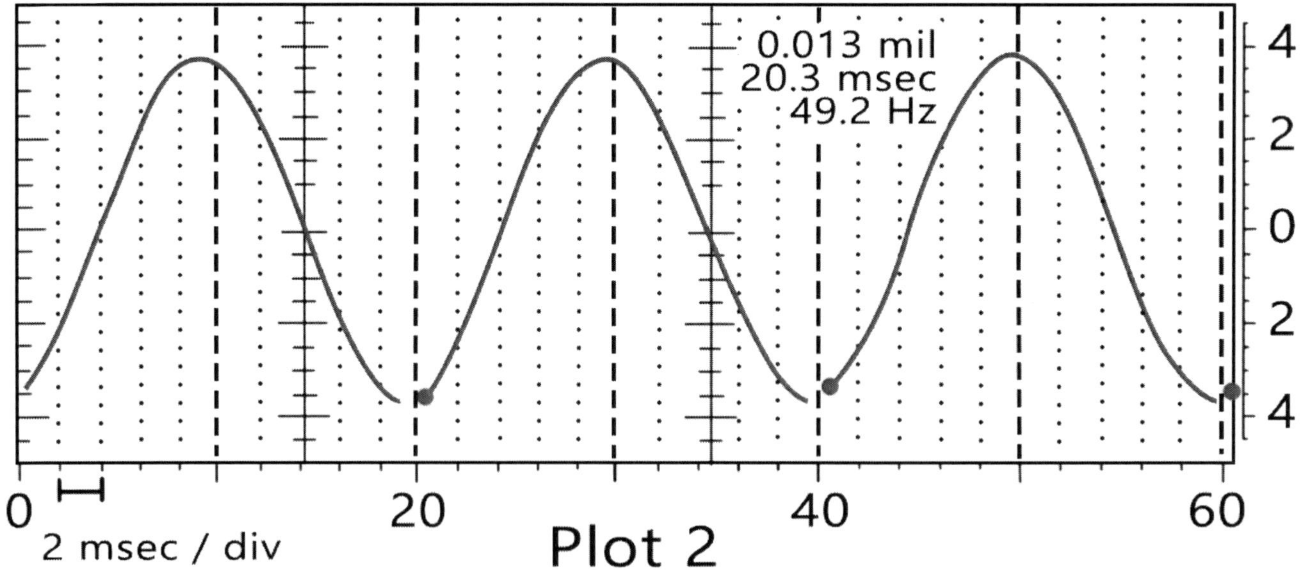

Plot 2

Fault Analysis and Correction

Question 067

Study the next vibration signal which is collected from a two pole motor. How many shaft turns are captured in this waveform?

A. 3 B. 20 C. 49.2 D. 50

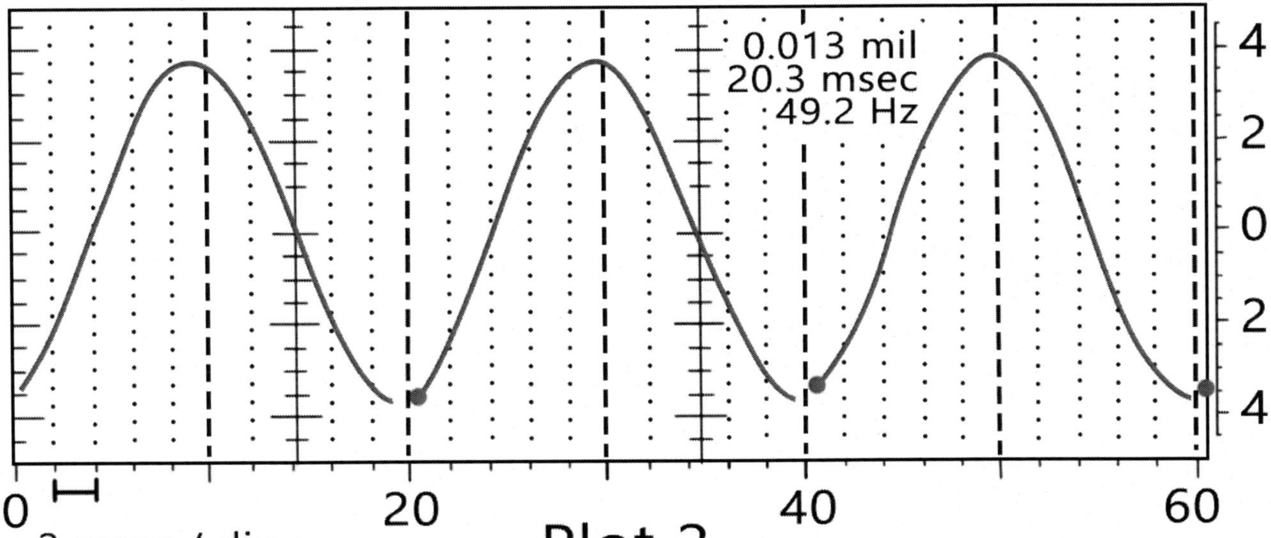

Plot 2

Fault Analysis and Correction

Question 068

Study the next vibration waveform collected from a two pole motor. How much time is captured in this vibration signal? How much time does the shaft take for a complete vibration cycle? How much time is needed to develop a complete shaft revolution?

	Time Captured	Vibration Cycle	Shaft Revolution
A.	2 msec	3 sec	3 sec
B.	20.3 msec	50 msec	60 msec
C.	49.2 sec	0.02 sec	20 msec
D.	0.06 sec	20 msec	0.02 sec

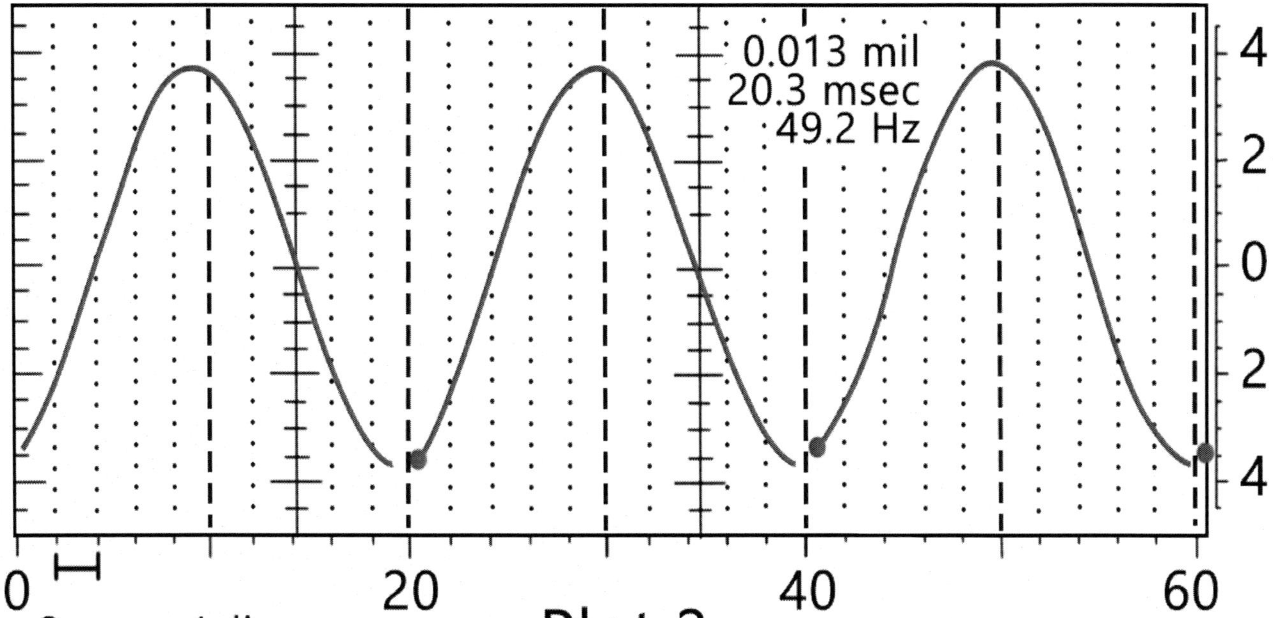

Plot 2

Note : ....

Fault Analysis and Correction

Question 069

What is the phase lag between the triggering signal and the vibration signal in time, parts of revolution and degrees? Refer to the next illustration.

	Time	Rotation	Degrees
A.	40 msec	1/4 turns	90°
B.	47 msec	3 turns	90°
C.	28 msec	3/4 turns	270°
D.	12 msec	3/4 turns	270°

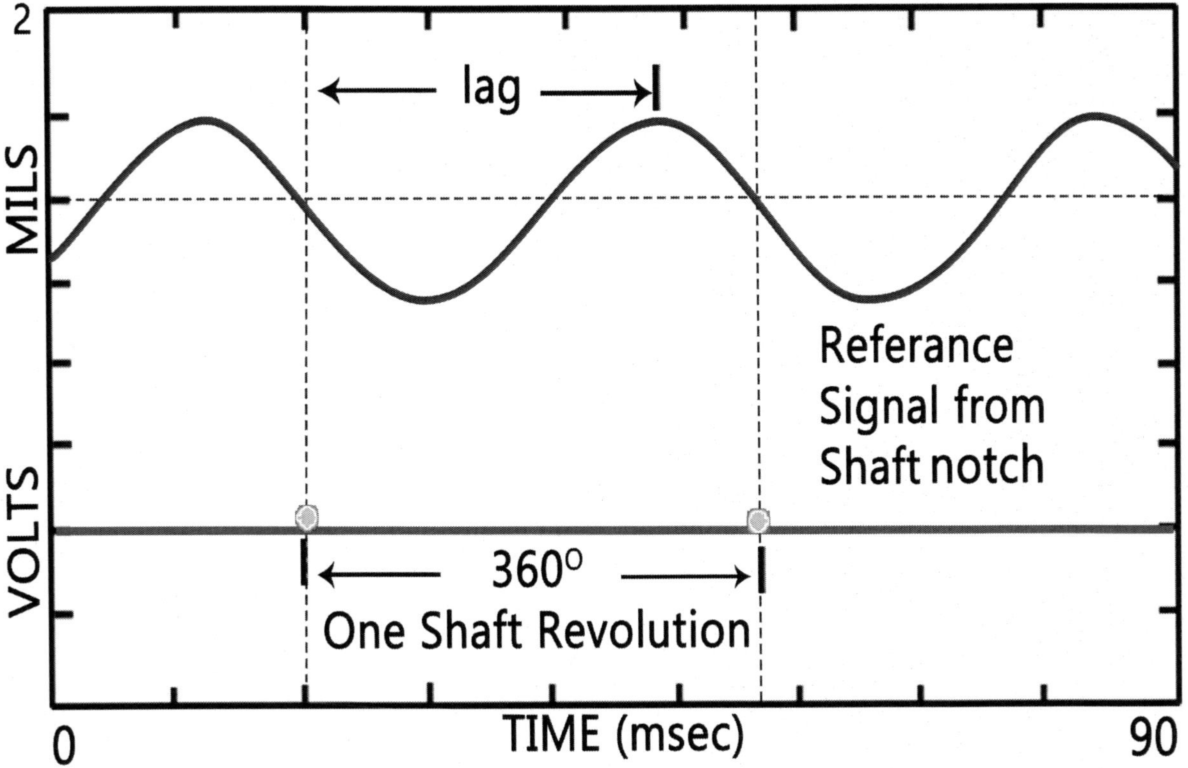
2
lag
MILS
VOLTS
Referance
Signal from
Shaft notch
360°
One Shaft Revolution
0
TIME (msec)
90

Note : ....

....

....

Fault Analysis and Correction

Question 070

What is the phase lag between the trigger signal and the vibration signal in degrees, parts of revolution and time? Refer to the next illustration.

	Degrees	Rotation	Time
A.	0°	1 turn	0 msec
B.	90°	1/4 turn	10 msec
C.	270°	3/4 turn	10 msec
D.	180°	½ turn	10 msec

Note : ....

....

....

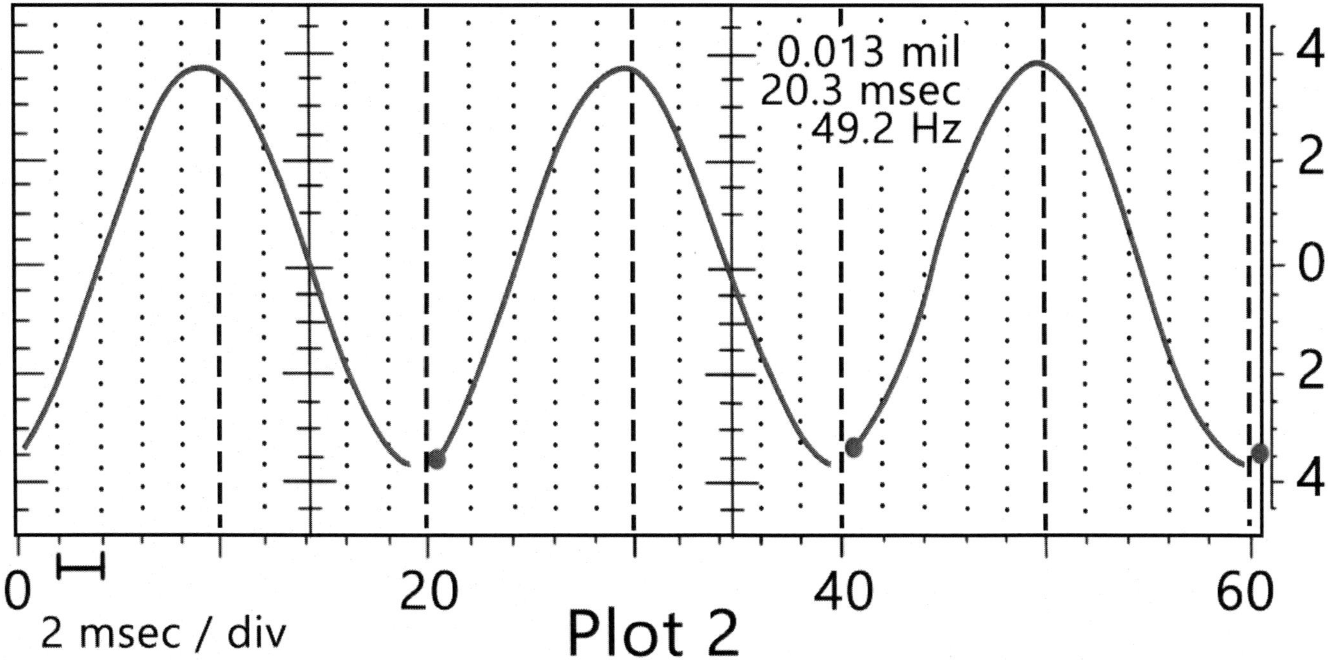

Plot 2

Note : ....

....

....

Fault Analysis and Correction

Question 071

For the next vibration waveforms, find the phase angle (lag) for Signal A and Signal B. Units are in degrees. Expect some approximations.

	A	B
A.	170	260
B.	260	170
C.	49.2	20.3
D.	180	125

Note : ....

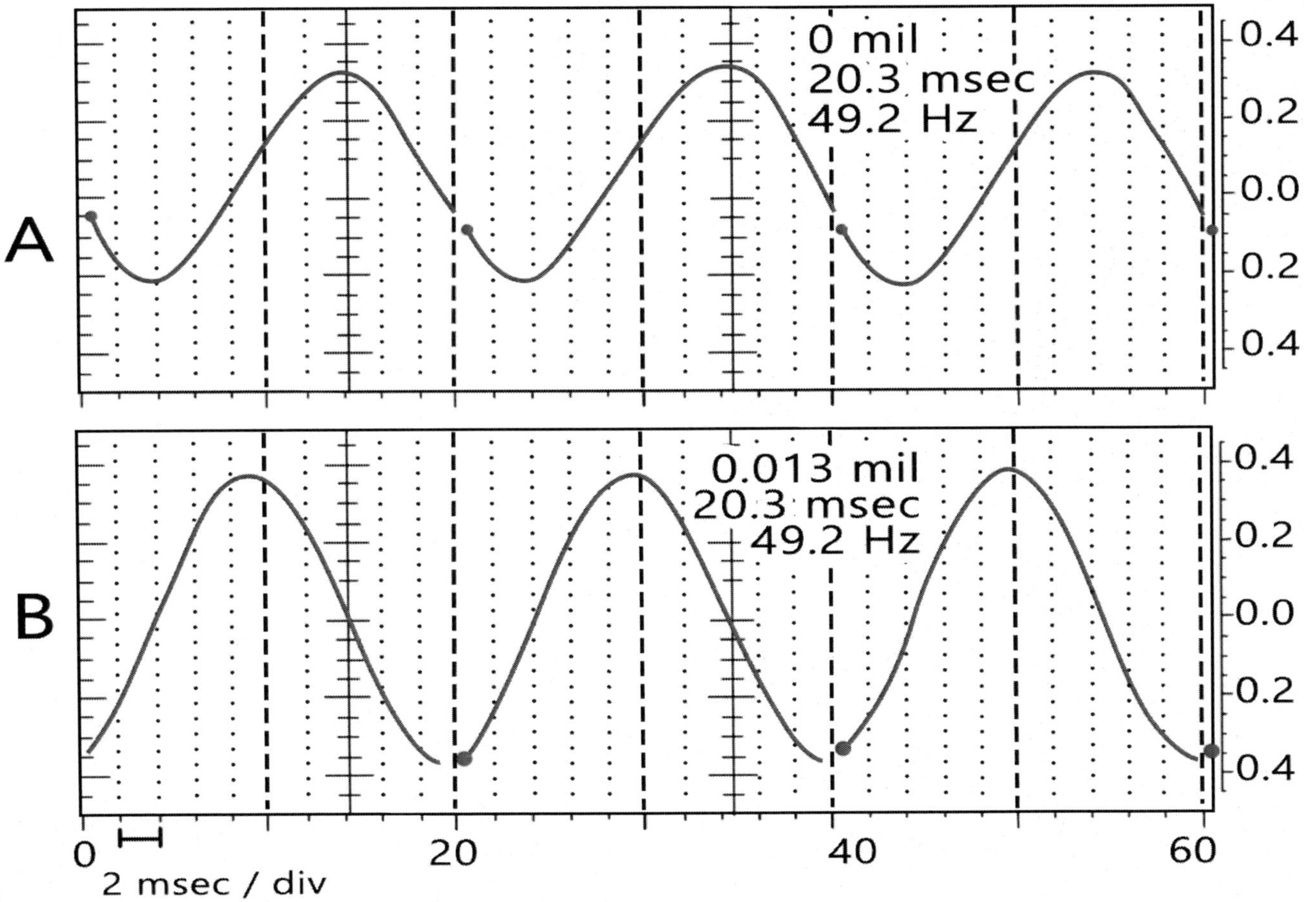
A
0 mil
20.3 msec
49.2 Hz
0.4
0.2
0.0
0.2
0.4
B
0.013 mil
20.3 msec
49.2 Hz
0.4
0.2
0.0
0.2
0.4
0
20
40
60
2 msec / div

Fault Analysis and Correction

Question 072

For the next plot, consider the zero time as a reference signal for the phase. What is the phase (in degrees) of the waveform?

A. Around 10

B. Around 90

C. Around 120

D. Around 360

Note : ....

....

....

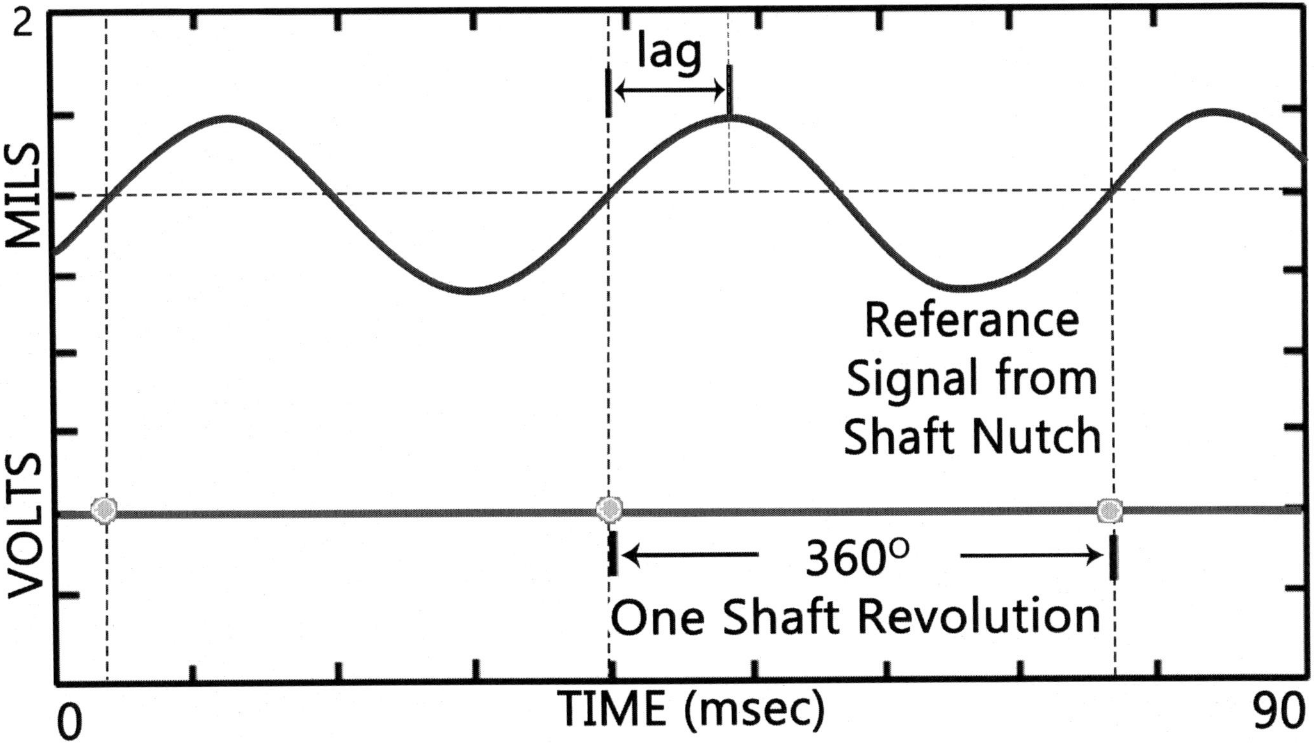
2
lag
MILS
VOLTS
Referance Signal from Shaft Nutch
360°
One Shaft Revolution
0
TIME (msec)
90

Fault Analysis and Correction

Question 073

Study to the next vibration waveform. What is the speed of the shaft? What is the vibration amplitude? Hint: the speed can be found through the key phasor signals.

	Speed	Amplitude
A.	20.3 rps	0.013 mil p-p
B.	No enough data is available	8 mil p-p
C.	2952 rpm	7.8 mil p-p
D.	60 Hz	4 mil p-p

Note : ....

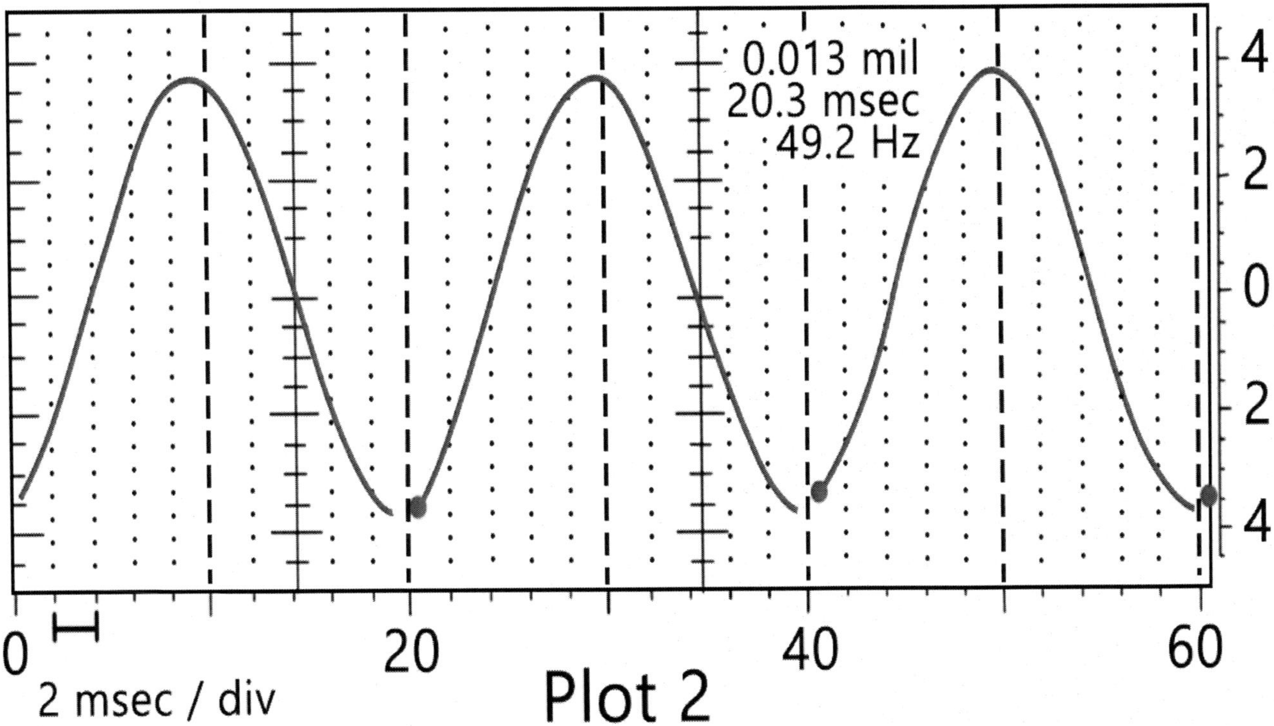

Plot 2

Fault Analysis and Correction

Question 074

For the next vibration waveforms, estimate the relative phase (in degrees) between Signal A and Signal B. Which signal leads? What is a good description for this case?

	Relative phase	Leading Signal	Description
A.	90	A leads	In phase
B.	180	B leads	Out of phase
C.	270	A leads	Opposite phase
D.	0	No signal leads	Similar phase

Note :

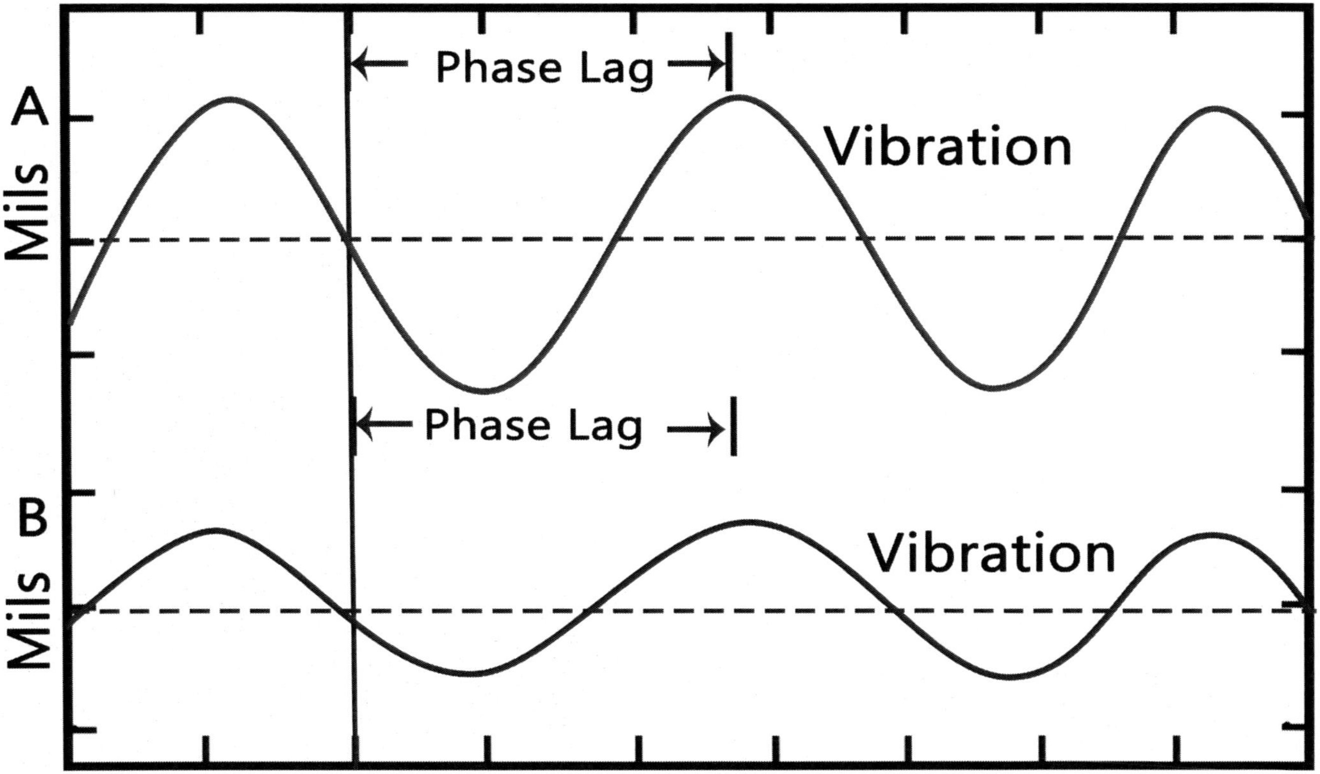
Phase Lag
A
Mils
Vibration
Phase Lag
B
Mils
Vibration

Fault Analysis and Correction

Question 075

For the next vibration waveforms, find the relative phase (in degrees) between Signal A and Signal B. Which signal leads? What is a good description for this case?

	Phase Difference	Leading Signal	Description
A.	90	A leads	In phase
B.	180	B leads	Out of phase
C.	180	No signal leads	Opposite phase
D.	360	No signal leads	Similar phase

Note : ....

....

....

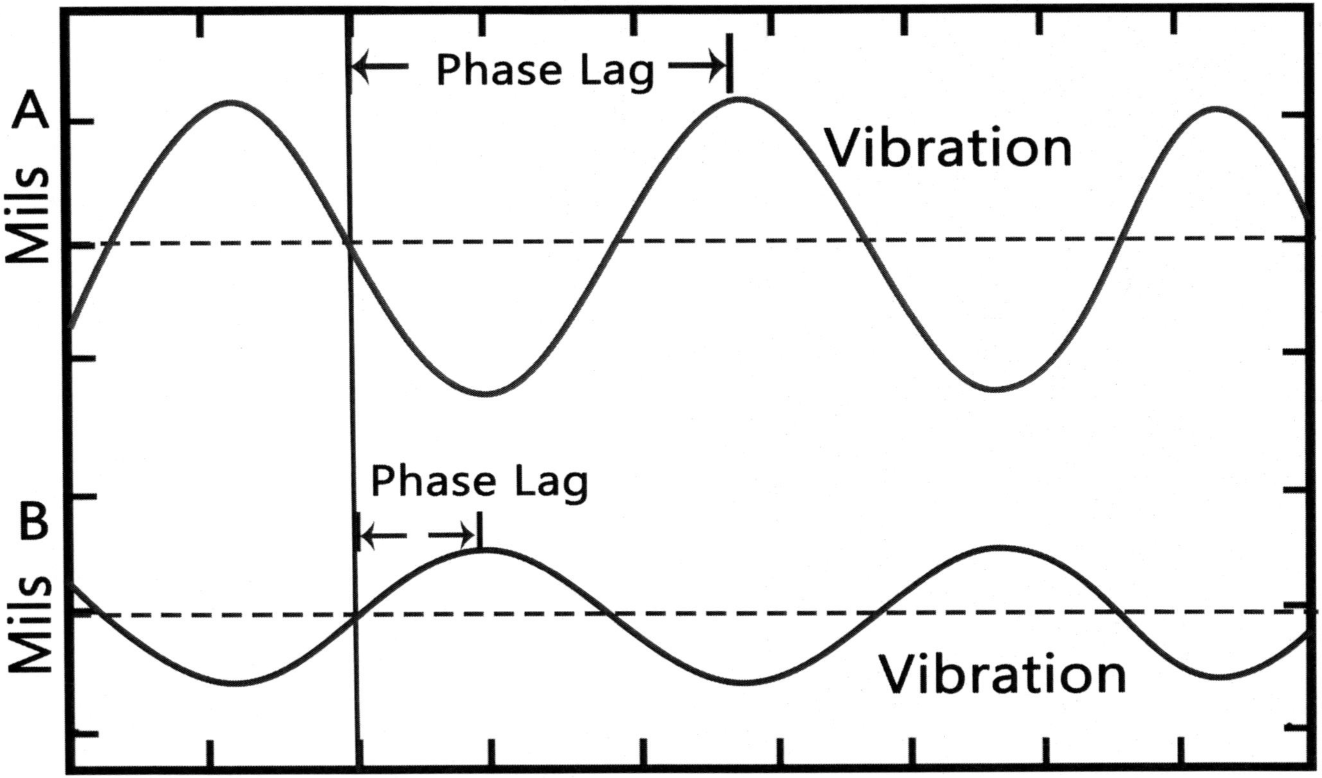

Note : ....

Fault Analysis and Correction

Question 076

For the next vibration waveforms, find the relative phase between Signal A and Signal B. Which signal leads? expect some rounding.

	Phase Lag (degree)	Leading Signal
A.	20.3	Signal A leads
B.	5	Signal A lags
C.	90	Signal B leads
D.	49.2	No signal leads

Note : ....

....

....

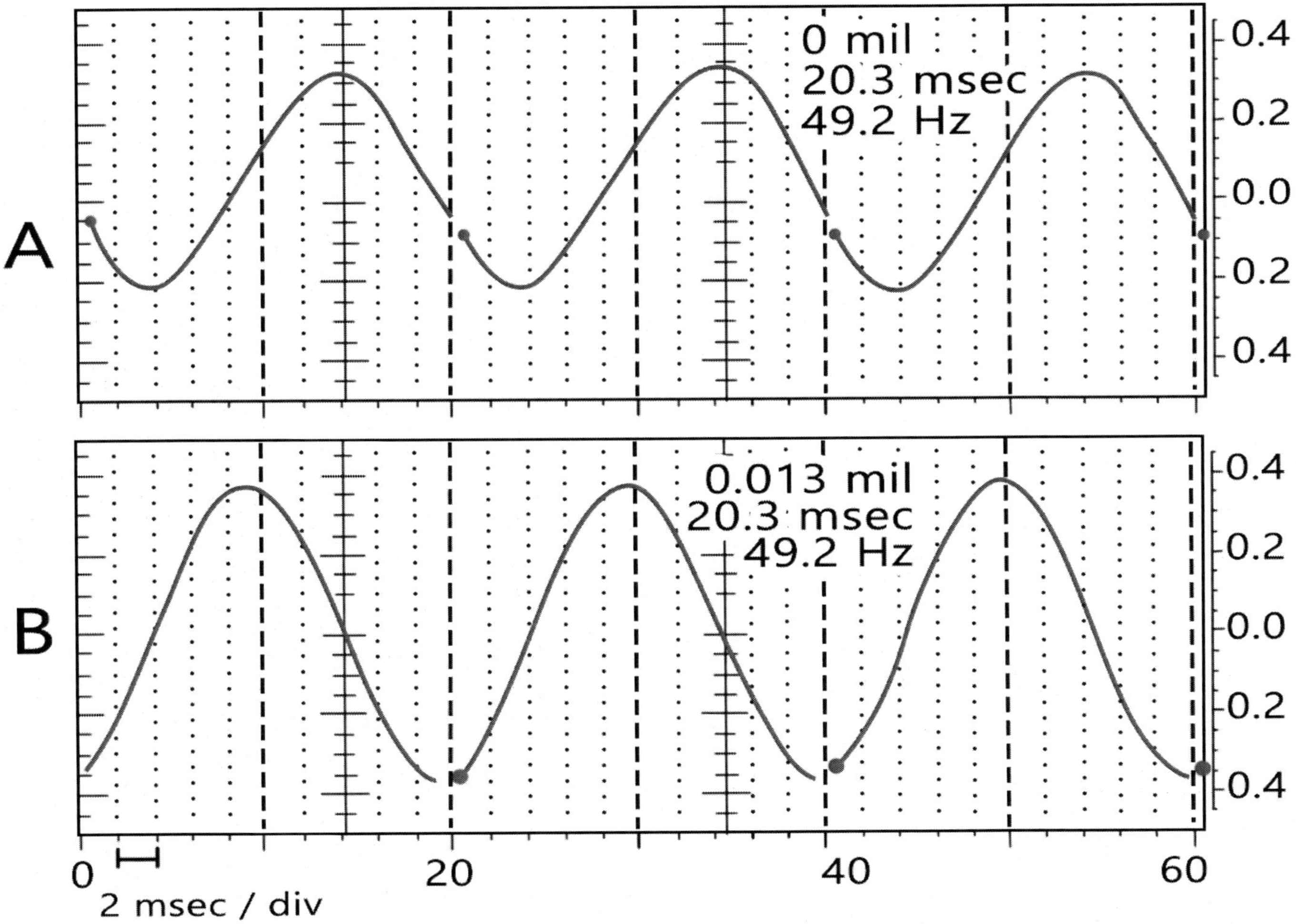
A
0 mil
20.3 msec
49.2 Hz
B
0.013 mil
20.3 msec
49.2 Hz
0.4
0.2
0.0
0.2
0.4
0
20
40
60
2 msec / div

Fault Analysis and Correction

Question 077

A vibration spectrum of a large motor (driving a pump) shows a very high amplitude exactly at 1X. The vibration is measured from the shaft which is mounted on journal bearings. What is a possible cause of the high vibration based on this information? Hint: Sketch the spectrum.

A. Rotor unbalance

B. Misalignment

C. Bent shaft

D. All of the above

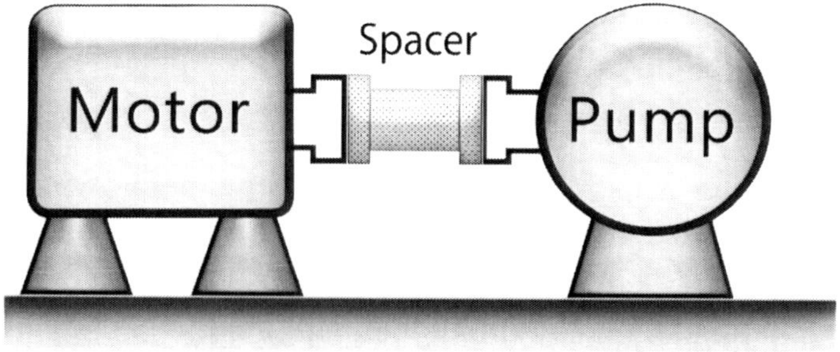

Note : ....

....

....

Fault Analysis and Correction

Question 078

What is the strongest vibration symptom for a severely damaged rolling element bearing? Select the best option from the below choices.

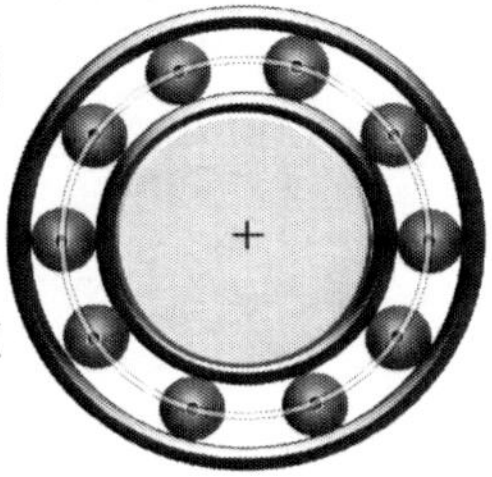

A. In a frequency spectrum, an analyst can see a major sinusoidal component at running speed.

B. In a frequency spectrum, an analyst can see a major component at 2 times running speed.

C. In a waveform, an analyst can see a major spike component at every other revolution.

D. In a waveform, an analyst can see several major spikes in each revolution.

Note : ....

....

....

Fault Analysis and Correction

Question 079

Which symptom is the best indication to recognize a defective ball bearing from its vibration spectrum? Refer to the next illustration.

A. Strong presence of non-synchronous components

B. Strong presence of synchronous components

C. Strong presence of 2LF components

D. Strong presence of 10X components

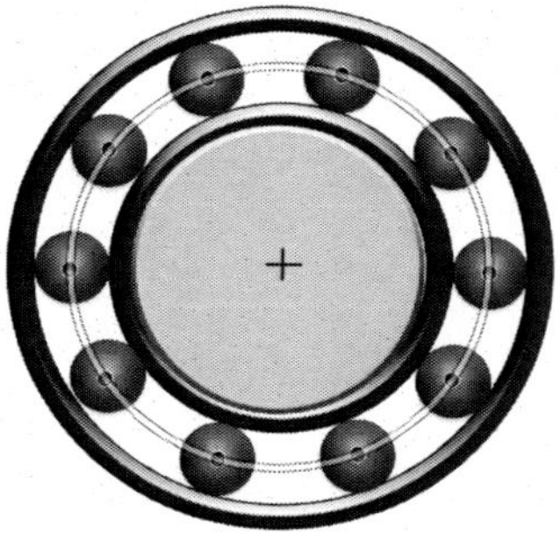

Note : ....

....

....

Fault Analysis and Correction

Question 080

For the next bearing, which of the following are the most possible inner race and outer race pass frequencies? Assume the shaft's speed is 1000 rpm. Hint: consider the typical characteristics of bearing fault frequencies, in terms of frequency orders. Then convert the frequency to other units.

	Inner	Outer
A.	1000 cpm	0 cpm
B.	1000 cpm	500 cpm
C.	6720 cpm	3280 cpm
D.	5000 cpm	10000 cpm

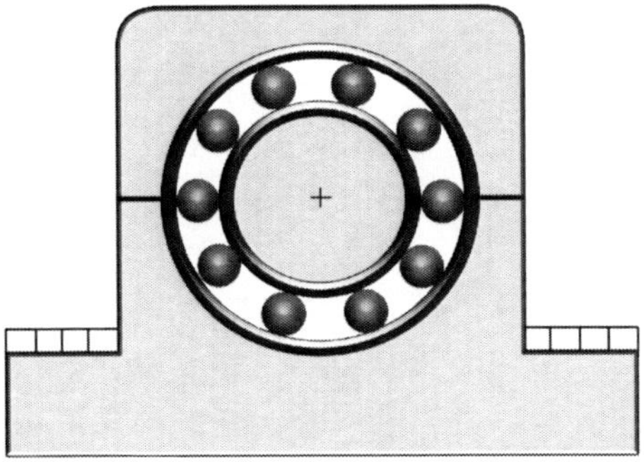

Fault Analysis and Correction

Question 081

What is a key vibration symptom that indicates a possible wear problem in a rolling element bearing?

A. Noisy waveform with clear non-integer spectral components.

B. Usually, the advanced stage of wear is associated with an audible noise and noticeable high frequency vibrations.

C. Fault components remain constant (amplitude and frequency) as the machine ramps up and down.

D. A and B

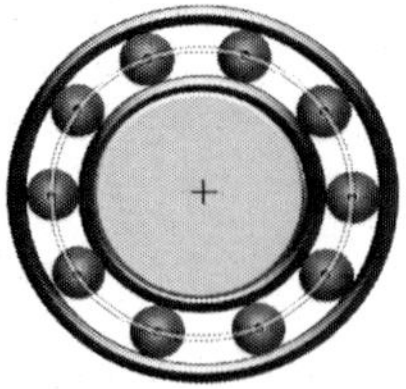

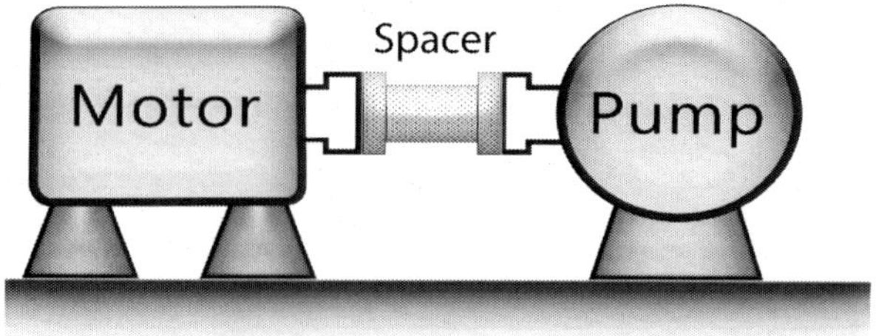

Fault Analysis and Correction
Question 082

In a vibration waveform plot, what is a classic symptom of a ball bearing damage? Refer to the next illustration.

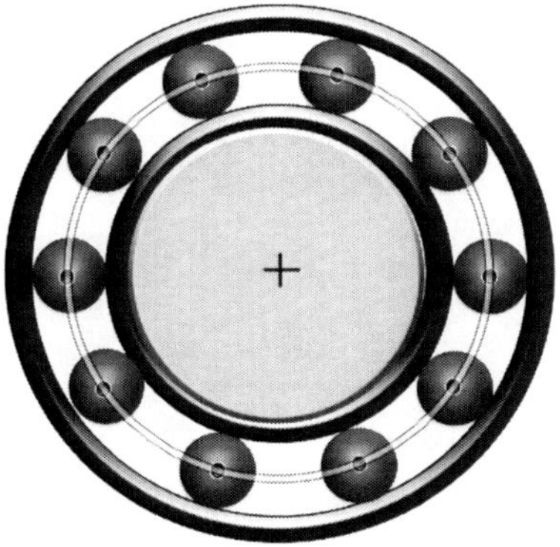

A. A strong 10X component especially if you collect a long waveform because this ball bearing has 10 balls.

B. Several sharp spikes in each shaft turn.

C. Pure sinusoidal wave especially if you collect high frequency data.

D. Square shape signal especially if you collect a phase angle trend.

Note : ......

Fault Analysis and Correction
Question 083

In a vibration waveform plot, what is a classic symptom of rotor unbalance for a motor driving a pump? Assume both motor and pump use ball bearings with 10 balls each.

A. A strong 10X component if you collect a long waveform

B. Several sharp spikes in each 1X cycle

C. Mostly a sinusoidal wave

D. Clean waveform with key component at the number of vanes in the impeller

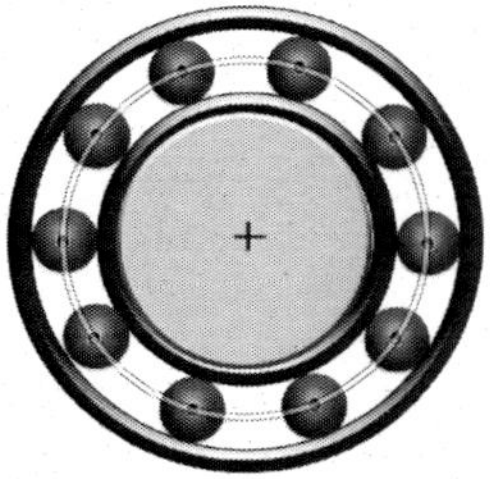

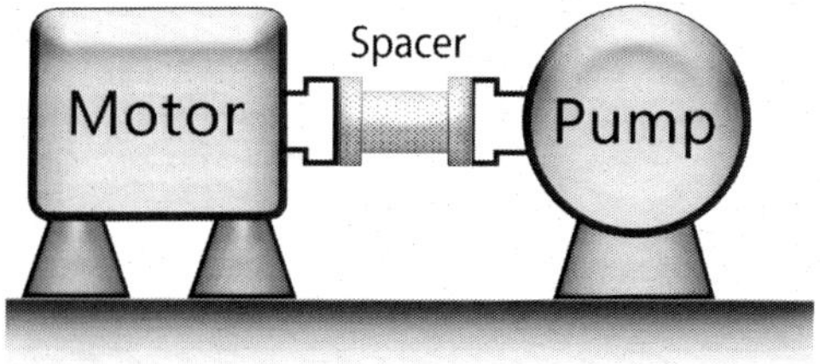

Note :

Fault Analysis and Correction

Question 084

For the next type of bearing, which of following are the most possible inner race and outer race pass frequencies? Assume the shaft's speed is 1800 rpm. Hint: consider the typical characteristics of bearing fault frequencies.

	Inner	Outer
A.	11340 cpm	6660 cpm
B.	4720 Hz	7668 Hz
C.	1800 cpm	0 cpm
D.	1800 cpm	900 cpm

Note :

Fault Analysis and Correction

Question 085

For the next type of bearing, which of the following is the most possible cage or fundamental pass frequency? Assume the shaft's speed is 1800 rpm. Hint: consider the typical characteristics of bearing fault frequencies in orders. Then, convert frequency to cpm or Hz.

A.	740 cpm
B.	740 Hz
C.	7400 cpm
D.	18000 cpm

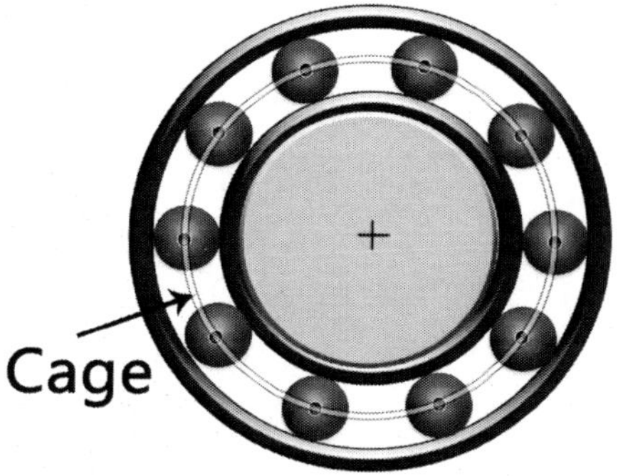

Note : ....

....

....

Fault Analysis and Correction

Question 086

A small fan (1498 rpm) is mounted on rolling element bearings. The vibration collected in velocity from the bearing housings shows several peaks separated by 25 Hz. What could be the fault? Refer to the next illustrations.

A. The information suggests a classic sign of looseness.

B. The information suggests a mature unbalance condition.

C. The information suggests a strong electrical grounding fault.

D. The information suggests an early bearing defect such as ball pittings.

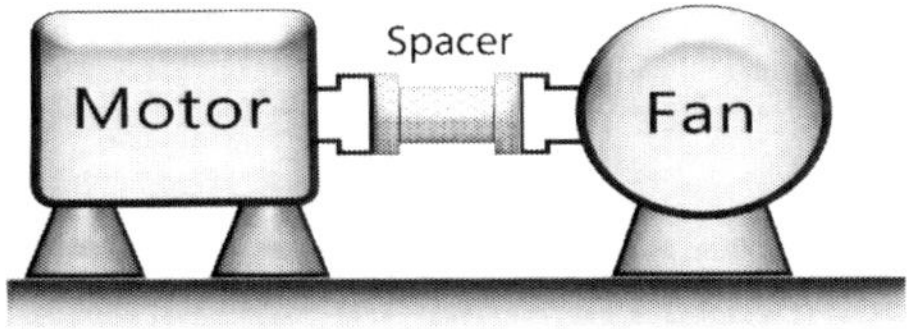

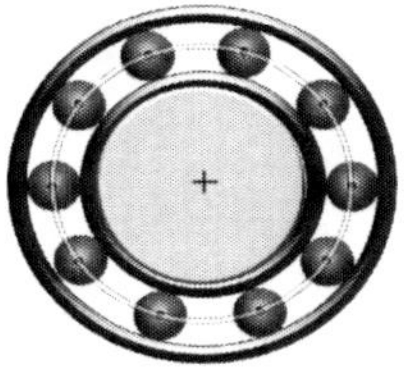

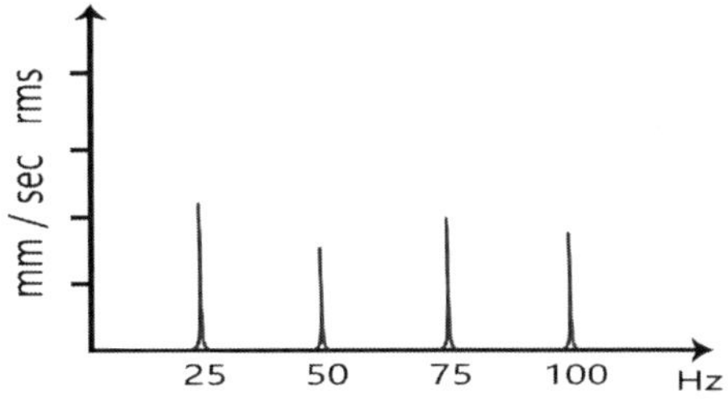

Fault Analysis and Correction

Question 087

In a vibration waveform plot, what is a classic symptom for a faulty vibration sensor/cable? Consider looseness in wire connections as a common cable fault. Refer to the next illustration.

A. A strong 10X component if bearings have 10 balls

B. Several sharp spikes in each 1X cycle

C. A pure gearmesh if you collect high frequency data

D. An unstable waveform shape and unstable overall amplitude.

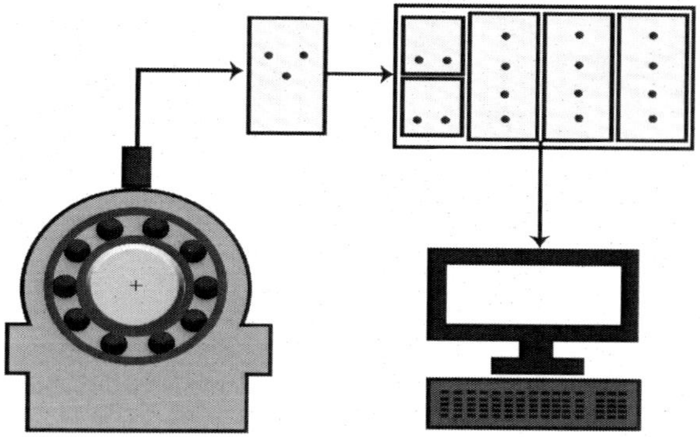

Fault Analysis and Correction
Question 088

This spectrum is for a fan (5 blades) mounted on rolling element bearings and running at about 2050 rpm. What is the most probable fault from the given choices?

A. Unbalance

B. Shaft bow

C. Looseness

D. Resonance

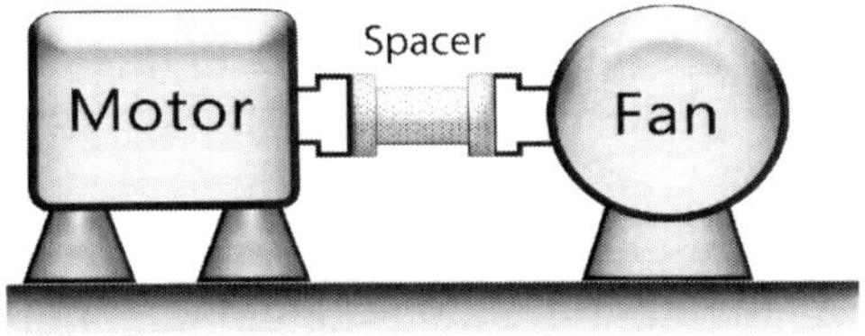

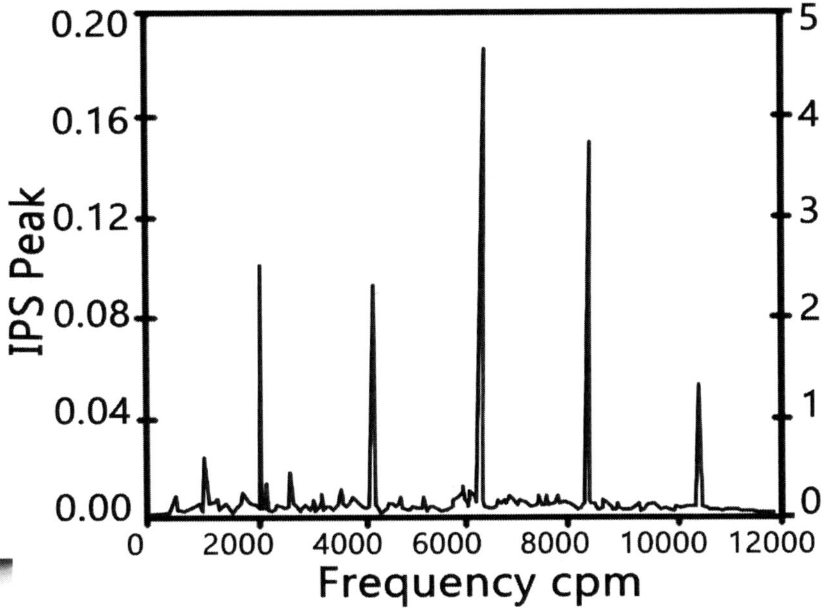

Fault Analysis and Correction

Question 089

A 60 Hz 4 pole induction motor is mounted on rolling element bearings. Each bearing has 8 balls. It is believed that the rotor has 35 bars while the stator has 53 slots. The motor drives a pump with 5 vanes. What are the typical frequencies (in cpm) of the following faults? Select the best option from the below choices. Hint: The speed of the motor and pump is slightly below 1800 rpm.

	Motor's Rotor mass Unbalance (cpm)	Shaft Misalignment at the Coupling (cpm)
A.	740	1765
B.	1498	9882
C.	1795	3590
D.	3590	7200

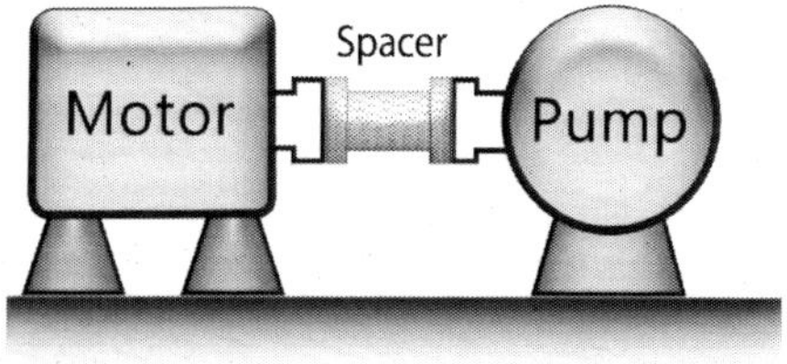

Fault Analysis and Correction

Question 090

Consider a simple fan driven by a motor. Which of the following vibration plots shows the strongest indication of a rotor unbalance? Hint: Recall the key symptoms of unbalance.

A. 1

B. 2

C. 3

D. 4

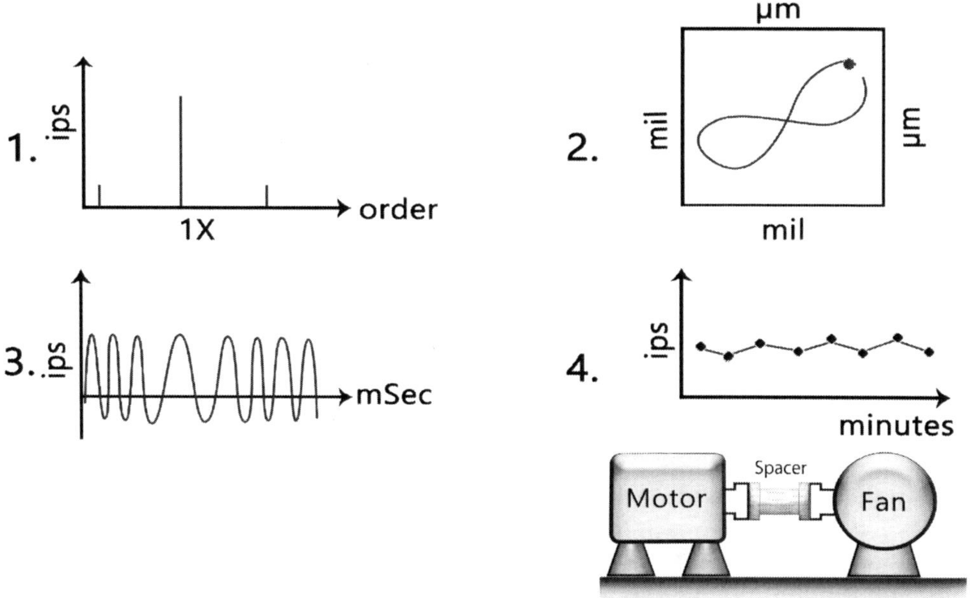

Fault Analysis and Correction

Question 091

Consider a simple fan driven by a horizontal motor. Which of the following vibration plots shows the strongest indication of an anti-friction bearing defect? Hint: Recall the key symptoms of a rolling element bearing fault.

A. 1

B. 2

C. 3

D. 4

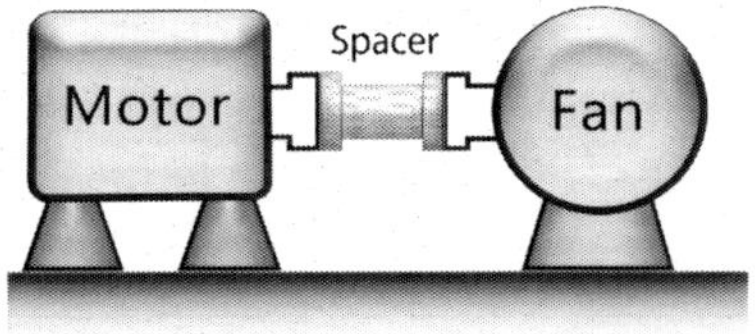

1.
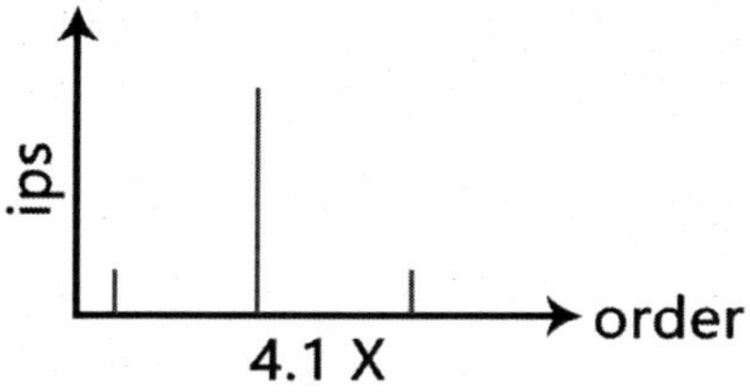

2.
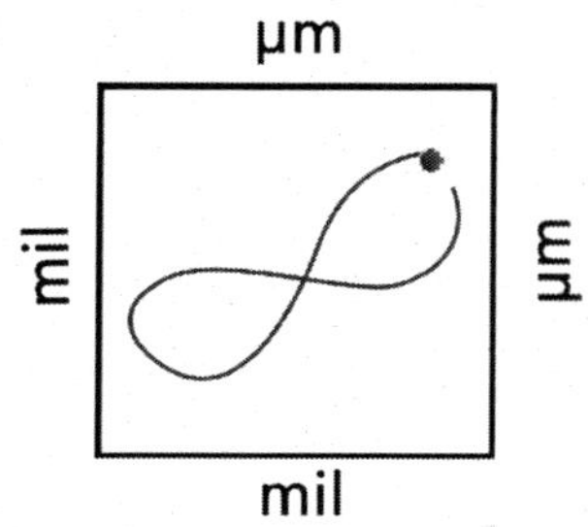

3.
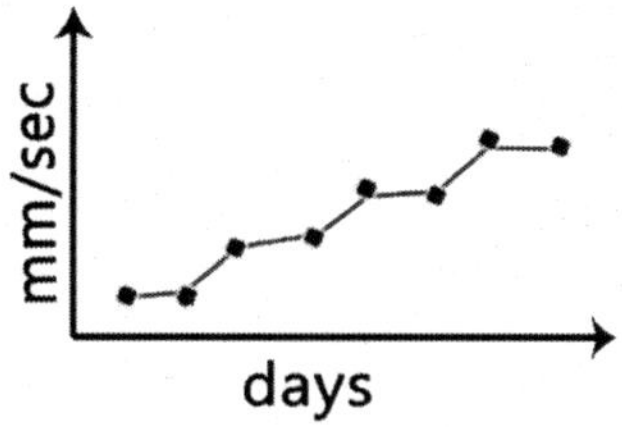

4.
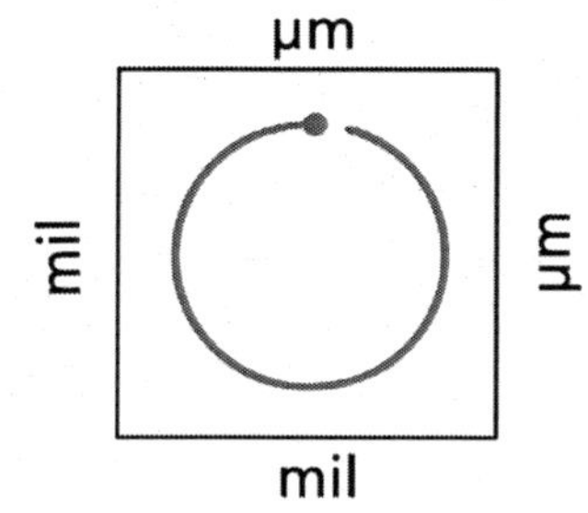

Fault Analysis and Correction

Question 092

Consider a simple fan driven by a motor. Which of the following vibration plots shows the most probable indication of a shaft misalignment?

A. 1

B. 2

C. 3

D. 4

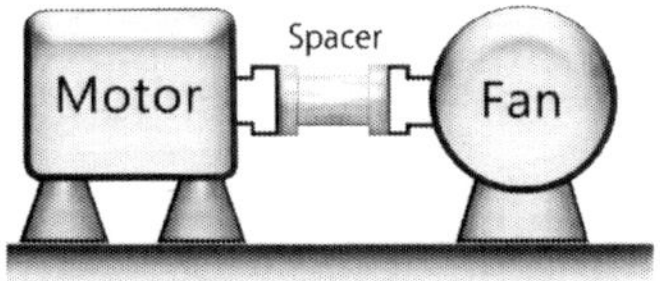

1.

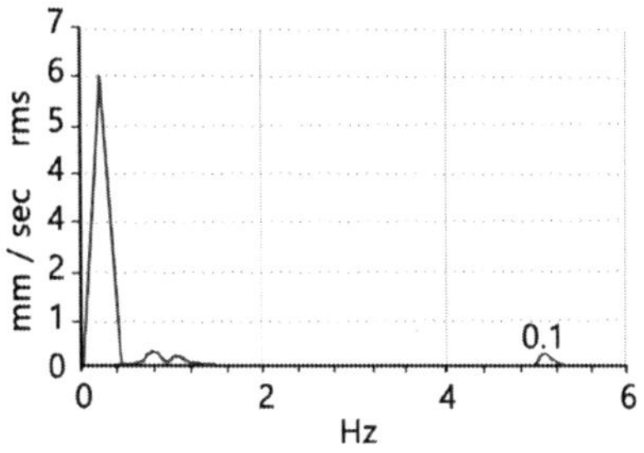

2.

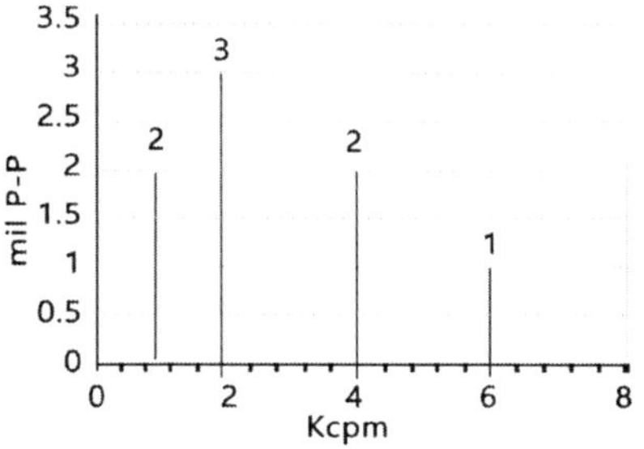

3.

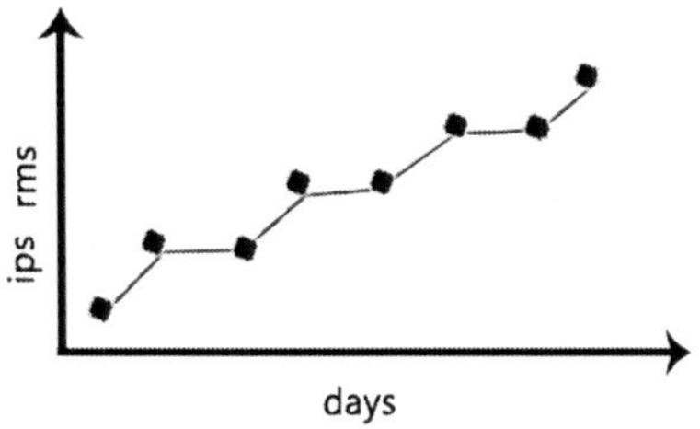

4.

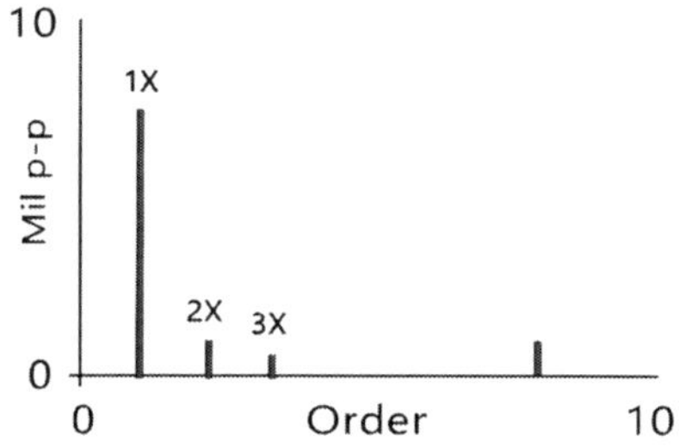

Fault Analysis and Correction

Question 093

Consider a simple fan driven by a vertical motor. Which of the following vibration plots shows the most probabale indication of a structural looseness?

A. 1

B. 2

C. 3

D. 4

1.

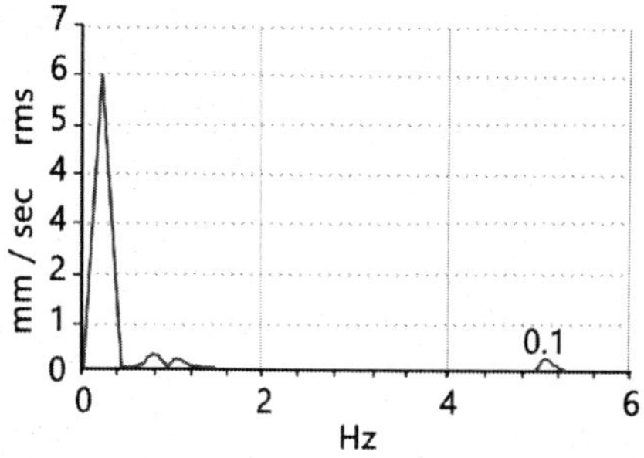

2.

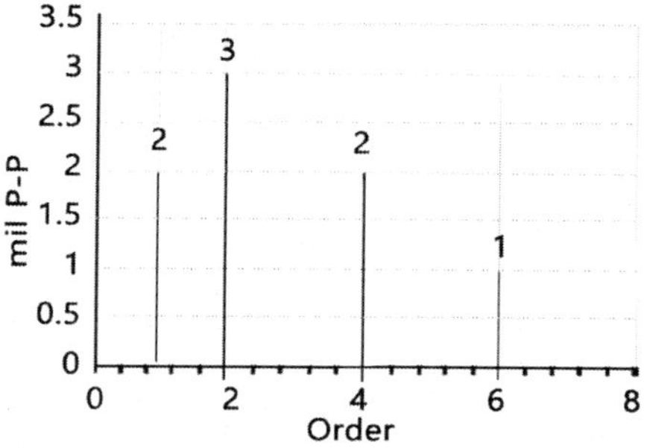

3.

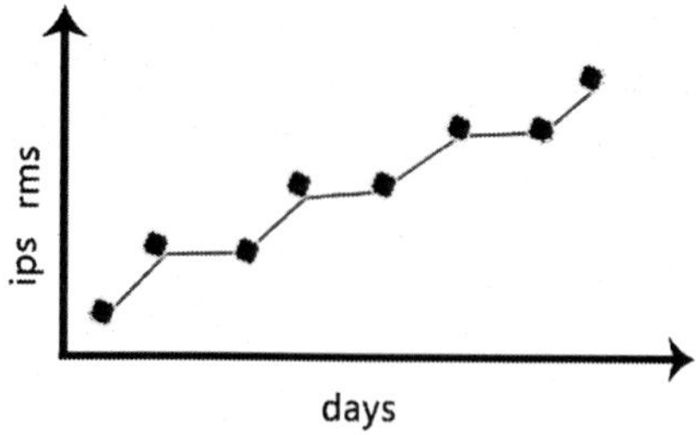

4.

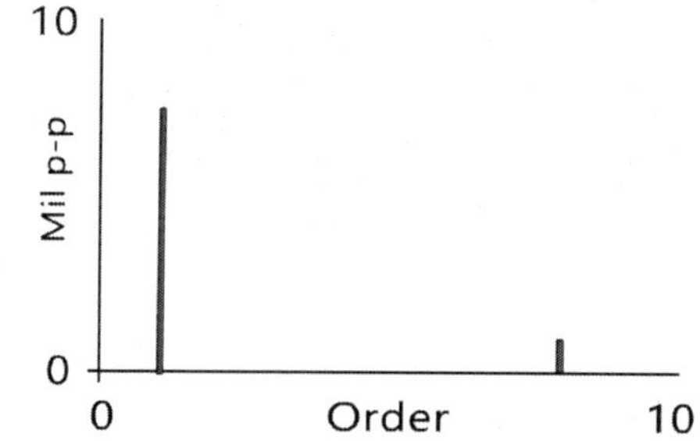

Fault Analysis and Correction

Question 094

Consider a simple fan driven by a motor. Which of the following vibration plots shows the most probable indication of a faulty vibration measurement loop?

A. 1

B. 2

C. 3

D. 4

1.

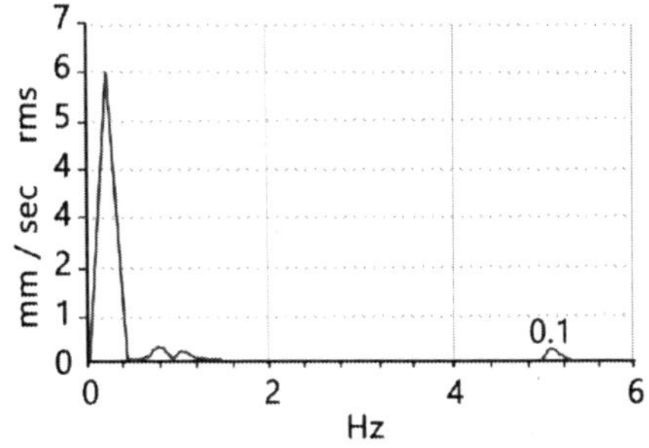

2.

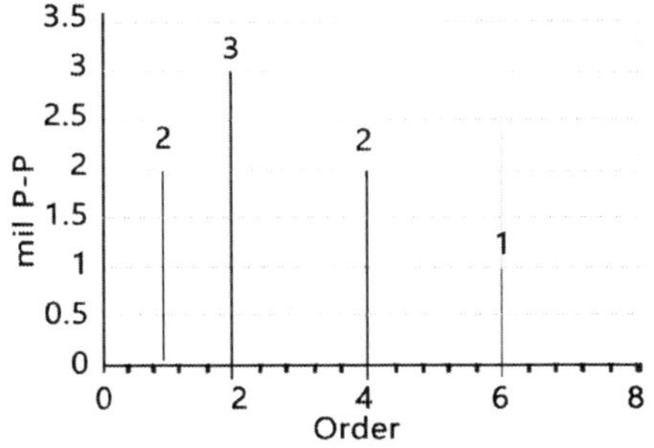

3.

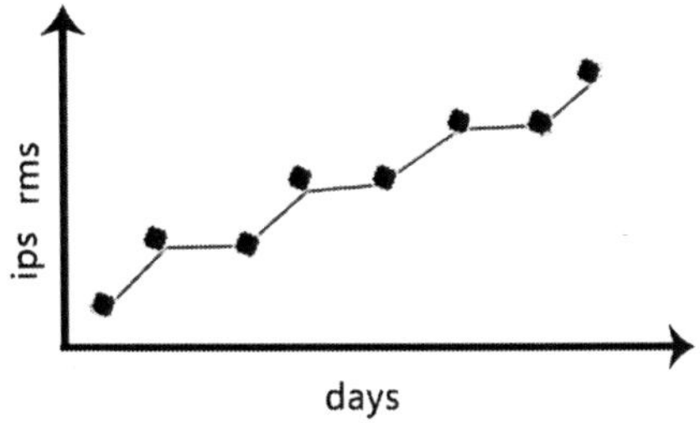

4.

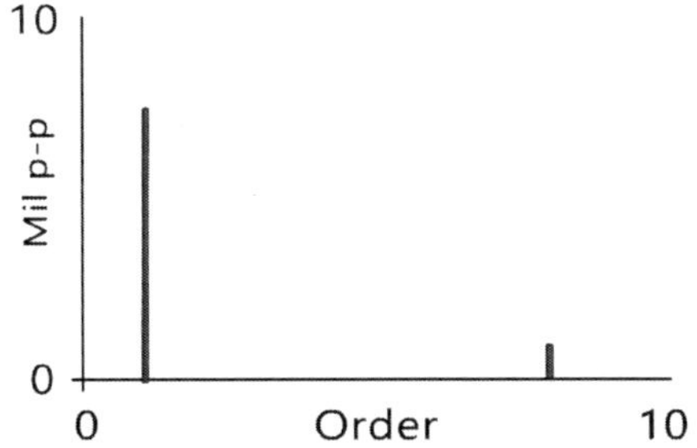

Fault Analysis and Correction

Question 095

Consider a 60 Hz 4 pole induction motor (running at 1790 rpm) mounted on rolling element bearings. Each bearing has 10 balls. It is believed that the rotor has 35 bars while the stator has 53 slots. What are the possible frequencies in cpm for the following faults? Select the best option from the below choices. Hint: some electrical faults are associated to the line frequency. A bearing cage defect generates a subsynchronous frequency.

	Electrical Related Problem (cpm)	Bearing Cage Defect (cpm)
A.	60	53
B.	53	35
C.	14400	1790
D.	7200	730

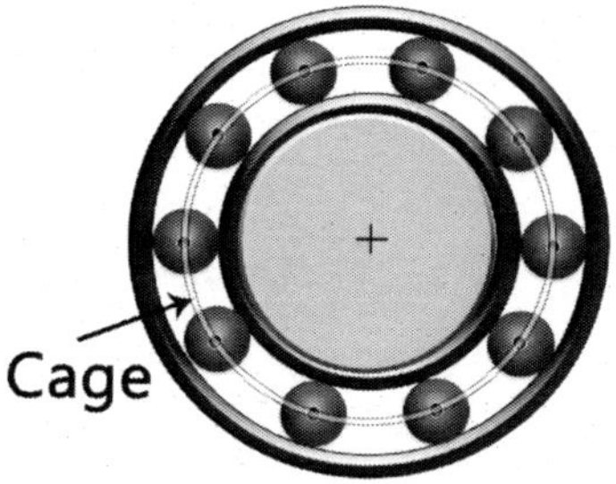

Fault Analysis and Correction

Question 096

Consider a 50 Hz 4 pole induction motor (running at 1490 rpm) is mounted on rolling element bearings. The bearing has 10 balls. It is believed that the rotor has 35 bars while the stator has 53 slots. What are typical frequencies in cpm for the following faults? Select the best option from the below choices. Hints: Some electrical faults are associated to the line frequency. A bearing cage defect generates a subsynchronous frequency.

	Electrical Related Fault (cpm)	Bearing Cage Fault (cpm)
A.	52150	615
B.	7200	149
C.	53	1500
D.	6000	12000

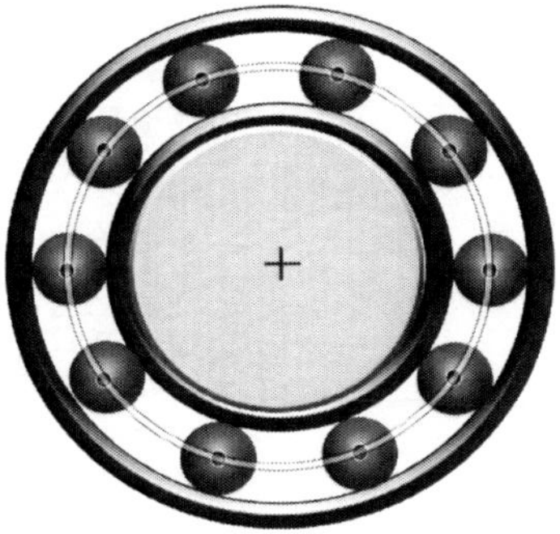

Fault Analysis and Correction

Question 097

Consider a 50 Hz 4 pole induction motor (running at 1495 rpm) driving a pump with a seven-vane impeller. Both driver and driven machines are mounted on rolling element bearings. What are typical frequencies in cpm of the following faults?

Select the best option from the below choices.

	Shaft Misalignment (cpm)	Vane Pass (cpm)
A.	1490	35000
B..	1500	7000
C.	2990	10465
D.	12000	7200

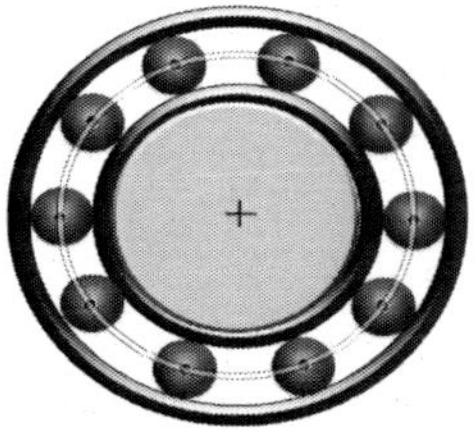

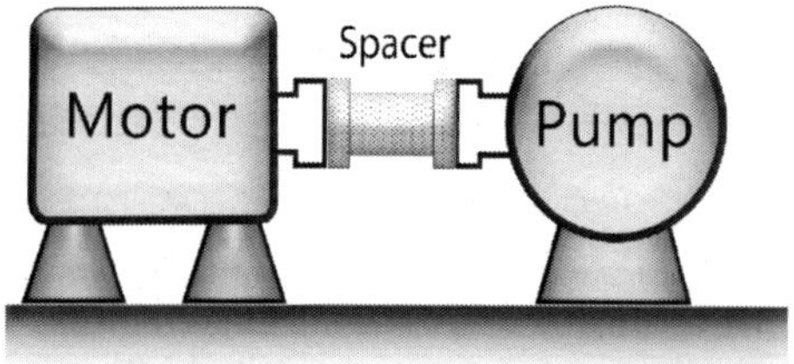

Fault Analysis and Correction

Question 098

Consider a vibration frequency spectrum of a horizontal electric motor (driving a pump). This spectrum shows a small 1X amplitude and a large 2X amplitude at the motor inboard bearing horizontal direction. All other measurement points show low and normal readings. What could be the most probable cause of this vibration based on the provided information? Hint: Sketch the spectrum.

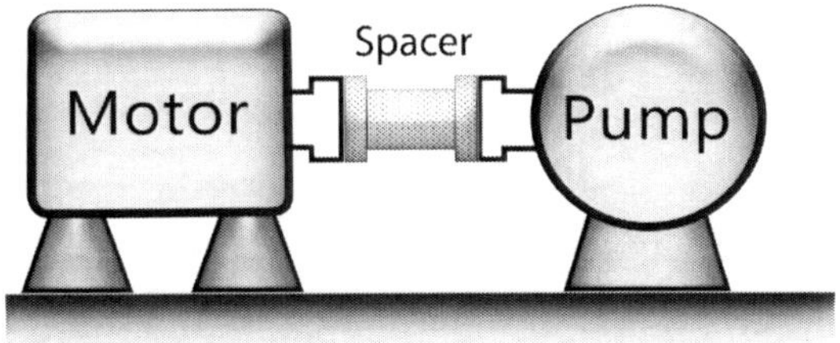

A. Shaft misalignment

B. Motor cooling fan unbalance due to a broken blade

C. Severe cavitation due to a high suction pressure (from pump side)

D. Bearing damage due to pitting in the motor's inboard bearing (inner race pass frequency)

Fault Analysis and Correction

Question 099

Small electric motors could have a relatively weak case construction. If a motor's case is distorted (while it is in service), what can be seen in the vibration spectrum at full speed? Hint: This problem causes changes in electromagnetic forces. An example of this case is a severe soft foot condition.

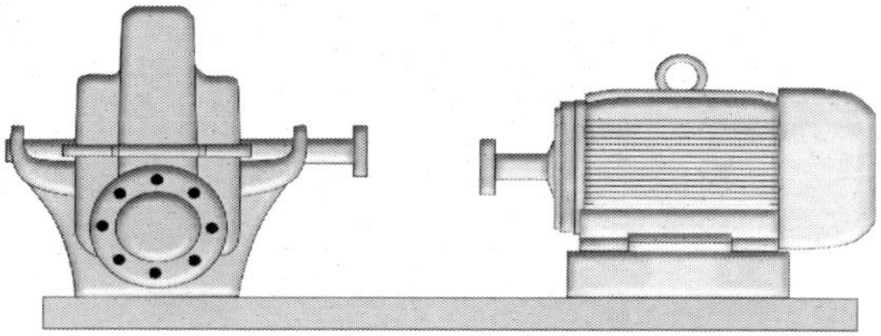

A. Peak at 2LF

B. Floor noise around 1X

C. High amplitude at 1/X frequency

D. High amplitude at rotor pass frequency

Note : ....

....

....

Fault Analysis and Correction

Question 100

Consider a vibration spectrum taken from a standard small horizontal induction motor designed to run at 1790 rpm. It runs with a high vibration amplitude exactly at 7200 cpm. What could be the fault of this case? This motor is in North America where the electricity is powered by 60 Hz. Hint: Sketch the spectrum.

A. An electrical fault

B. Shaft misalignment

C. Rotor unbalance

D. Bearing housing looseness

Note : ....

Fault Analysis and Correction

Question 101

Study the next nameplate which is for a small induction motor. What is the slip absolute frequency? How much is the slip as a ratio? Hint: Find running speed and synch speed then calculate the slip.

	Absolute Slip	Slip Ratio
A.	1.9 cpm	0.19
B.	20 rpm	0.0067
C.	50 Hz	1.94
D.	0.15 %	19

Serial	1L6DH4-SE41		
TYPE	TTXYZ	FARME	286T
H.P.	30.00	Service	1.15
AMPS	34.9	VOLTS	460
R.P.M.	2980	HERTZ	50
DUTY	CONT 40° AMB		

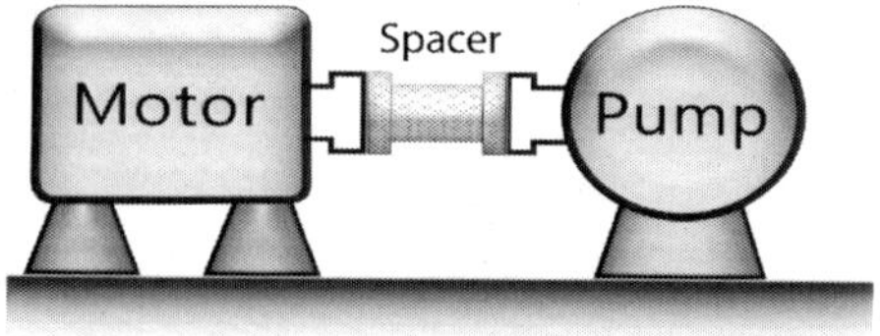

Fault Analysis and Correction

Question 102

Study the next nameplate which is for a small induction motor driving a pump with 5 vanes. What is the synchronous speed? How much is the slip as a percentage? What is the mass unbalance frequency? Hint: Slip is the speed difference between the running shaft speed and the synchronous speed.

	Synchronous Speed	Slip %	Unbalance (cpm)
A.	30 rpm	0	60
B.	30 Hz	0.28	1795
C.	1795 rpm	1.98	30
D.	600 Hz	100	149.6

Serial	1L6DH-4SE41		
TYPE	TTXYZ	FARME	286T
H.P.	30.0	Service	1.15
AMPS	34.9	VOLTS	460
R.P.M.	1795	HERTZ	60
DUTY	CONT 40° AMB		

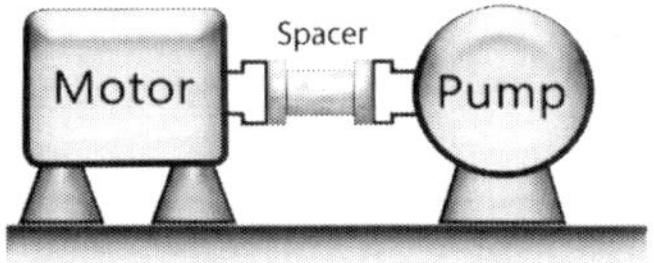

Fault Analysis and Correction

Question 103

Study the next nameplate which is for a small motor. What is the synchronous speed? If the driven pump has 6 vanes, what is the vane pass frequency? Hint: VPF = V * Speed.

	Synchronous	VPF
A.	20 rpm	209.4 Hz
B.	50 Hz	300 Hz
C.	2980 cpm	18000 cpm
D.	3000 rpm	17880 cpm

Serial	1L6DH4-SE41		
TYPE	TTXYZ	FARME	286T
H.P.	30.00	Service	1.15
AMPS	34.9	VOLTS	460
R.P.M.	2980	HERTZ	50
DUTY	CONT 40° AMB		

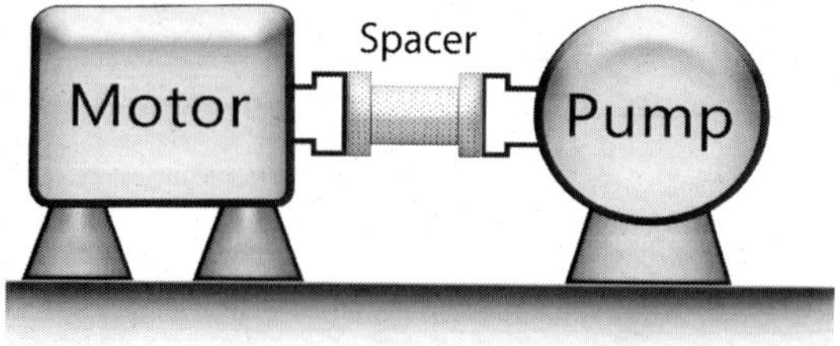

Fault Analysis and Correction

Question 104

Study the next name plate which is for a small motor. What is the mass unbalance frequency in cpm? What is the absolute frequency of slip? Hint: Slip is the speed difference between the running shaft speed and the synchronous speed.

	Unbalance (cpm)	Slip
A.	20	34.9 Amp
B.	1500	35 rpm
C.	3000	20 Hz
D.	2980	0.33 Hz

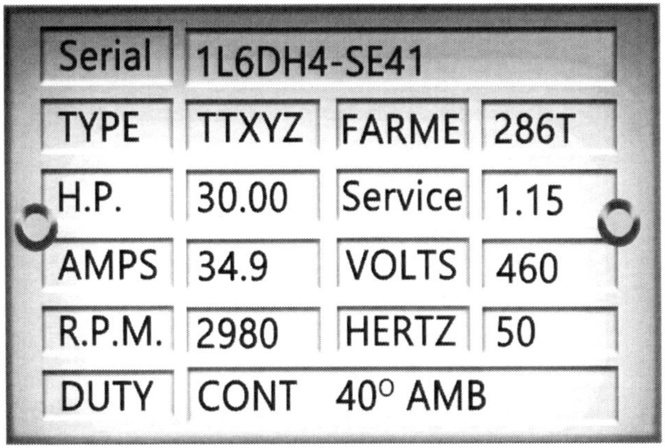

Fault Analysis and Correction

Question 105

Consider a standard induction motor (60 Hp or 45 kW) running at 1495 rpm. How many poles are in this motor? What is the electric line frequency in this case?
Hint: Use the equation for calculating the synchronous speed based on the poles and line frequency: NS= 2x LF ÷ Poles.

	Poles	LF
A.	1 poles	25 Hz
B.	2 poles	30 Hz
C.	3 poles	60 Hz
D.	4 poles	50 Hz

Fault Analysis and Correction

Question 106

Consider a standard induction motor (67 Hp or 50 kW) running at 1785 rpm. How many poles are in this motor? What is the power line frequency in this case?
Hint: Use the equation for calculating the synchronous speed based on the poles and line frequency: NS= 2x LF ÷ Poles

	Poles	LF
A.	1 poles	25 Hz
B.	2 poles	30 Hz
C.	3 poles	50 Hz
D.	4 poles	60 Hz

Fault Analysis and Correction

Question 107

Consider a standard induction motor (6 Hp or 4.5 kW) running at 1798 rpm. This motor drives a horizontal blower. What is the mass unbalance frequency, slip and the preferred vibration amplitude unit to use in this case? Hint: The preferred unit of vibration amplitude depends the type of bearing and the targeted frequency range.

	Unbalance Frequency	Slip	Preferred Amplitude Unit
A.	60 Hz	2 Hz	mil
B.	1795 cpm	60 Hp	mil
C.	1805 rpm	3%	ips
D.	30 Hz	2 cpm	ips

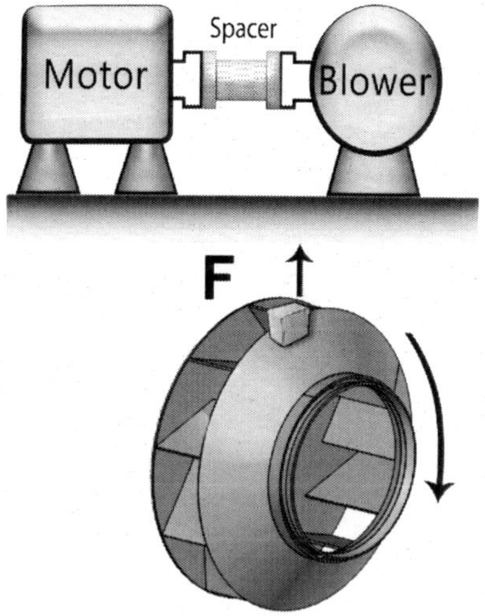

Fault Analysis and Correction

Question 108

A new small size motor runs at 995 rpm, when it is at its full load. This induction motor experiences a case distortion which leads to a 2LF vibration. What would be the order of this fault when it is detected by a vibration system? Hint: Identify the LF from the information in the question, then find the 2LF in cpm and as an order.

	Fault Frequency
A.	2.00 X
B.	4.00 X
C.	4.12 X
D.	6.03 X

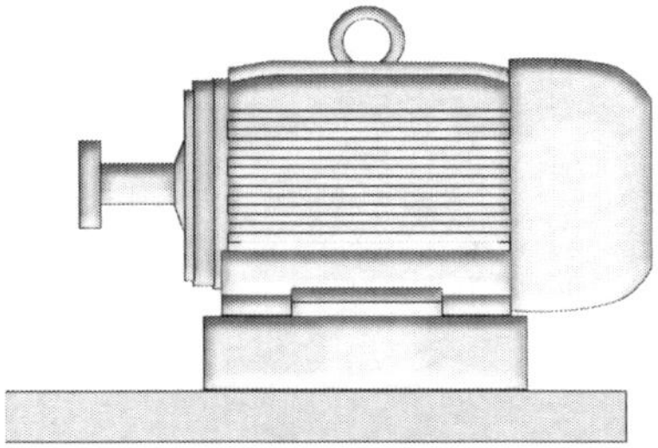

Fault Analysis and Correction

Question 109

A new small size motor runs at 888 rpm, when it is at its full load. This induction motor experiences a case distortion, which leads to a 2LF vibration. What would be the order of this fault when it is detected by a vibration system? What is the slip? Hint: Identify the LF from the information in the question, then find the 2 LF in cpm and order.

	Fault Order	Slip (%)
A.	2.00 X	8
B.	4.00 X	12
C.	4.12 X	120
D.	8.11X	1.3

Fault Analysis and Correction

Question 110

This spectrum is for a horizontal synchronous motor running at 3599.8 rpm at the time of data collection. The motor drives compressor. What could be the most probable fault? Slect from the given choices.

A. Shaft Misalignment

B. Looseness

C. Unbalance

D. All of the above

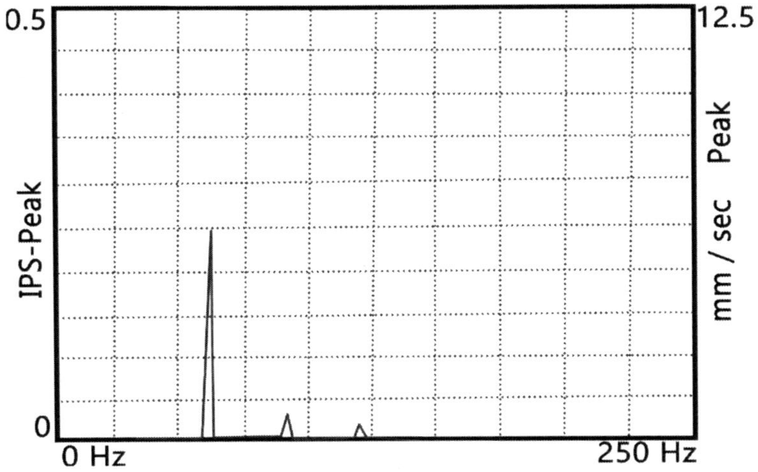

Note : ....

....

....

Fault Analysis and Correction

Question 111

Based on the nameplate and the waveform plot, what is the most probable fault in this small size motor? Hint: find the frequency and order of the dominant signal from the waveform then calculate the order. Next, compare the calculated order with the order of each of the given answers.

A. Rotor unbalance

B. Shaft misalignment

C. Bearing defect

D. Faulty vibration instrument circuit

Serial	1L6DH4-SE41		
TYPE	TTXYZ	FARME	286T
H.P.	30.00	Service	1.15
AMPS	34.9	VOLTS	460
R.P.M.	3595	HERTZ	60
DUTY	CONT	40°C AMB	

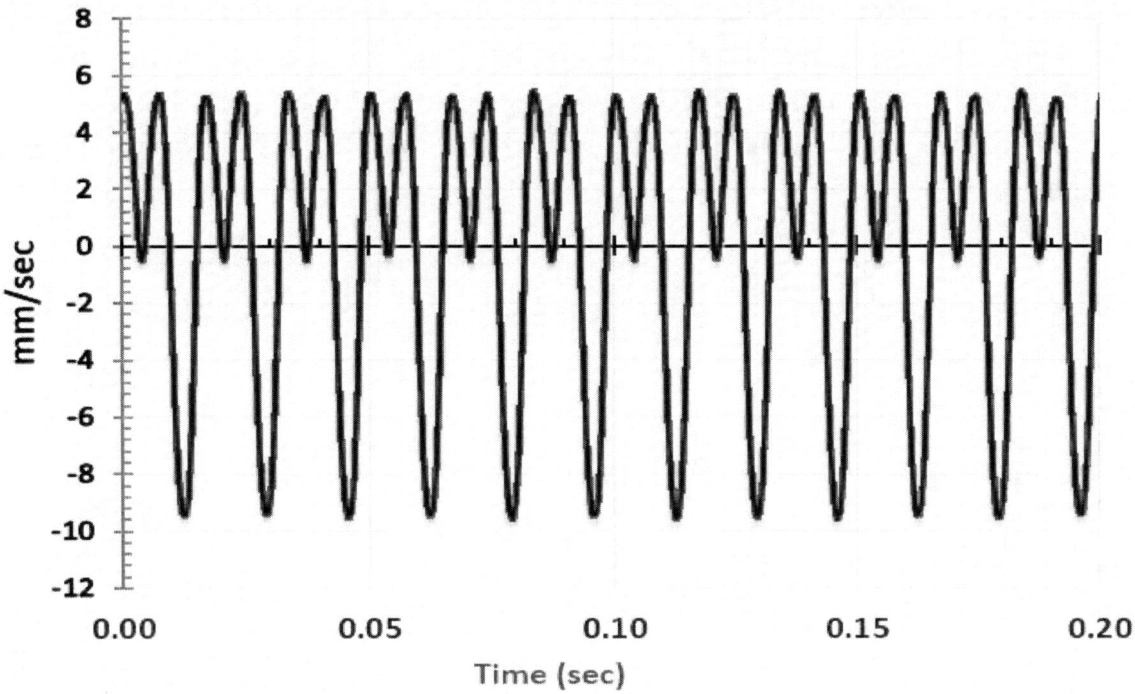
8
6
4
2
0
-2
-4
-6
-8
-10
-12
mm/sec
0.00
0.05
0.10
0.15
0.20
Time (sec)

Fault Analysis and Correction

Question 112

Based on the nameplate and the waveform plot, what is the most probable fault in this small size motor? Hint: find the frequency of the dominant signal from the waveform then calculate the order. Next, compare the calculated order with the order of each of the given faults.

A. Rotor unbalance

B. Shaft misalignment

C. Bearing defect

D. Faulty vibration instrument circuit

Serial	1L6DH4-SE41		
TYPE	TTXYZ	FARME	286T
H.P.	30.00	Service	1.15
AMPS	34.9	VOLTS	460
R.P.M.	1488	HERTZ	50
DUTY	CONT	40°C AMB	

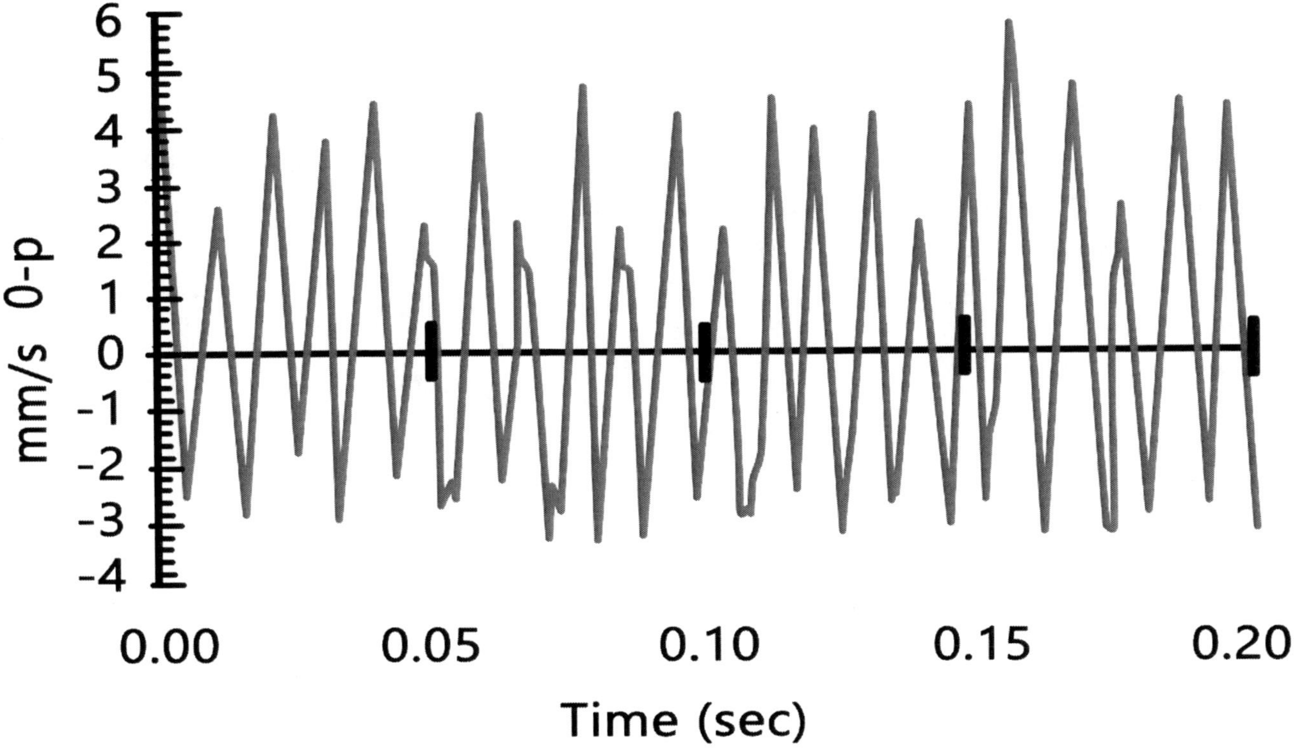
6
5
4
3
2
1
0
-1
-2
-3
-4
mm/s 0-p
0.00
0.05
0.10
0.15
0.20
Time (sec)

Fault Analysis and Correction

Question 113

Based on the nameplate and the frequency spectrum, what is the most probable fault in this motor? The measurement is taken from motor outboard bearing. Hint: Find the order of each peak in the spectrum. Then compare them to the orders of each fault from the answers. All of the given faults generate synchronous components except bearing defects.

A. Rotor unbalance

B. Shaft misalignment

C. Bearing defect

D. Looseness

Serial	1L6DH4-SE41		
TYPE	TTXYZ	FARME	286T
H.P.	30.00	Service	1.15
AMPS	34.9	VOLTS	460
R.P.M.	3595	HERTZ	60
DUTY	CONT	40° C AMB	

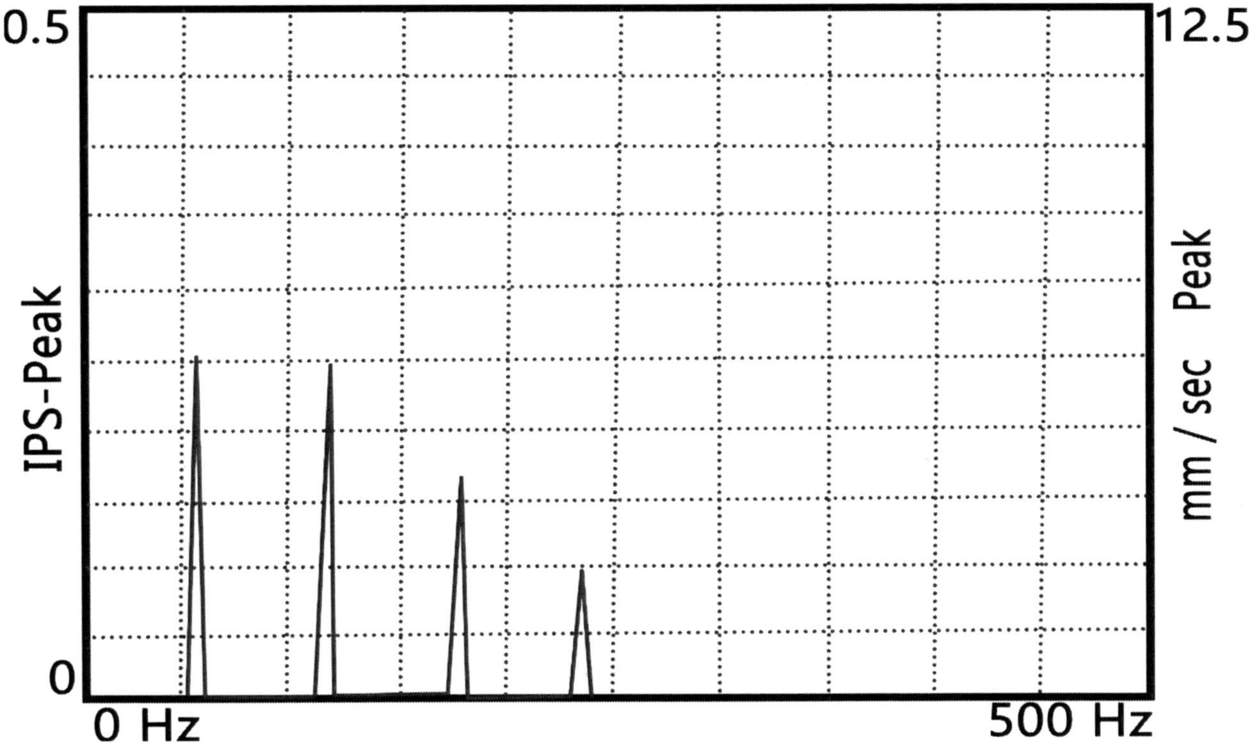
0.5
0
IPS-Peak
12.5
mm / sec Peak
0 Hz
500 Hz

Fault Analysis and Correction

Question 114

If this impeller rotates at 52300 rpm, what is its vane pass fault frequency? This type of impeller is commonly used in plant air compressors. Hint: count the vanes.

A. 15690 Hz or 18X

B. 15690 cpm or 18X

C. 52300 Hz or 17X

D. 52300 cpm or 19X

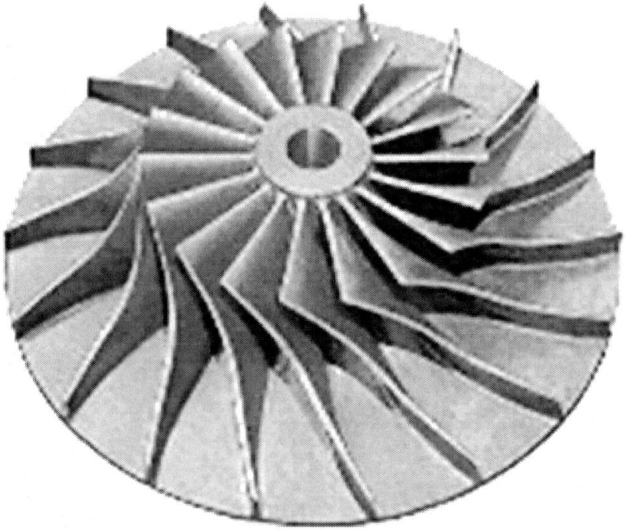

Fault Analysis and Correction

Question 115

Consider a 5-blade fan experiencing 12.5 mm/sec 0-p (0.5 ips 0-p) vibration. The fan runs at 1490 rpm. If the problem is unbalance, what is the vibration frequency order in this case?

A. 1X

B. 5X

C. 1490 X

D. 62.5 mm/sec 0-p

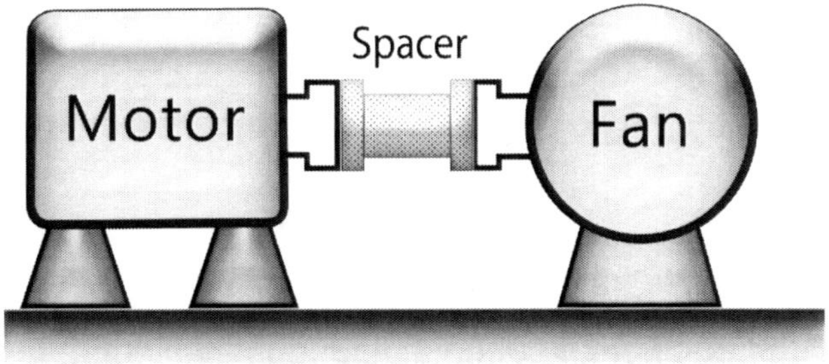

Note : ....

....

....

Fault Analysis and Correction

Question 116

This type of impellers is commonly used in plant air compressors. This impeller rotates at 52300 rpm. What is its most probable vibration frequency if there is a random dirt collected on the vanes?

A. 15690 Hz or 18X

B. 15690 cpm or 18X

C. 52300 Hz or 17X

D. 52300 cpm or 1X

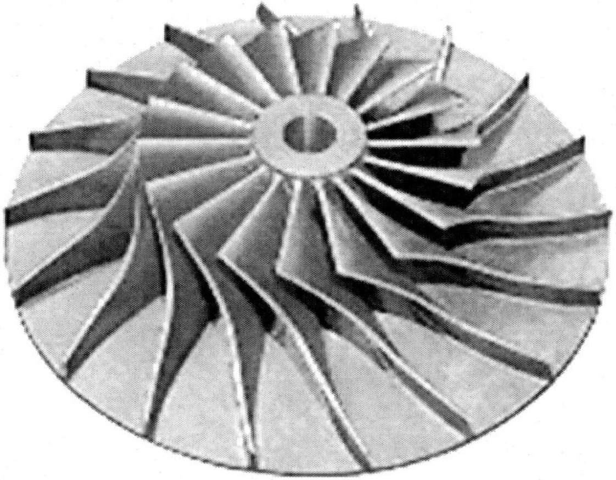

Note : ....

....

....

Fault Analysis and Correction

Question 117

Consider a 9-blade fan experiencing 12.5 mm/sec 0-p (or 0.5 ips 0-p) vibration. The fan runs at an unknown speed. If the problem is unbalance, what is the vibration frequency in cpm and order in this case?

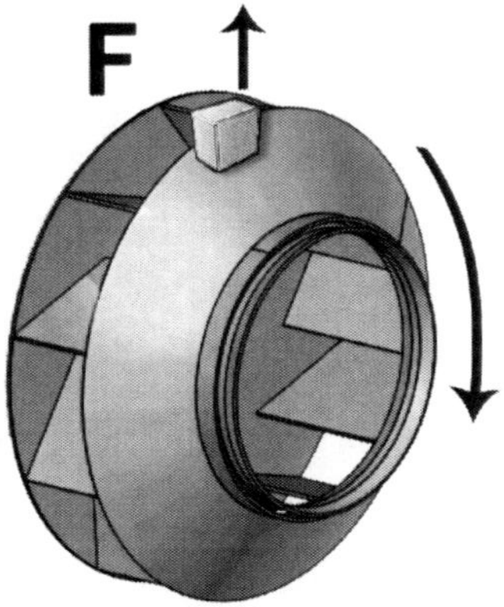

A. 1495 cpm,1X

B. 1795 cpm, 1X

C. 2995 cpm, Unknown order

D. Unknown frequency, 1X

Note : ....

....

....

Fault Analysis and Correction

Question 118

What is true about "sidebands" in the field of machinery vibration analysis? Hint: sidebands are commonly found in gearbox vibrations.

A. They are the second and third most dominant frequency components in a gearbox vibration.

B. They are the second and third harmonics of the fundamental frequency.

C. They are spectral amplitudes (usually small) located on both sides around a key vibration amplitude.

D. They reflect a frequency range defined by the vibration analyst and it is used to establish band alarms.

Fault Analysis and Correction

Question 119

For a single-stage speed-decreasing gearbox, how many gear mesh frequencies are calculated? Refer to the next illustration.

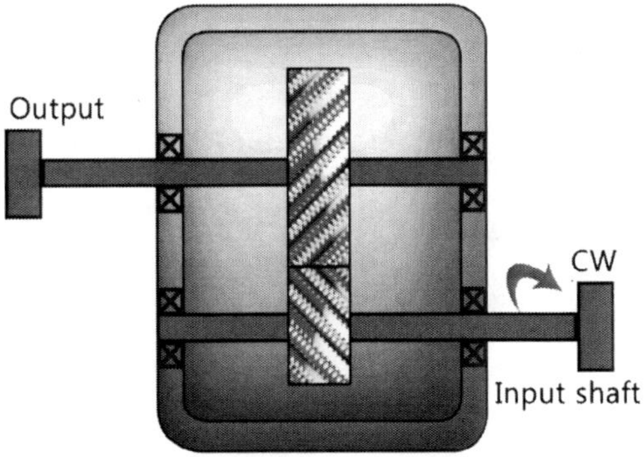

A. One for the entire gearbox.

B. Two. Each shaft generates a unique mesh frequency.

C. Three, one for each shaft and the third is the combination of both.

D. The correct answer depends on the speed ratio of the input and output shafts. A gear mesh frequency is generated every time the speed is doubled.

Fault Analysis and Correction

Question 120

What is a common vibration symptom that indicates a severe gearbox fault?

A. High GMF and its harmonics.

B. Presence of relatively high (noticeable in the linear scale) sidebands around the GMF.

C. High 3X accompanied by continuous phase change at 0.44 X.

D. A and B.

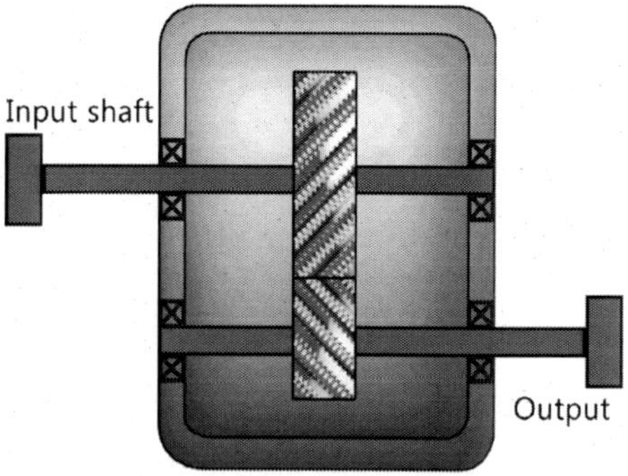

Fault Analysis and Correction

Question 121

Study the next simplified illustration for a gearbox. Input speed is 1500 rpm. Which statement is true?

A. The speed on output shaft is higher than that of the input shaft.

B. The direction of rotation of the output shaft is CCW (using a common viewing convension).

C. This gearbox has two (2) gear mesh frequencies.

D. To detect the high frequency faults of this gearbox, F_{max} must be <10,000 cpm.

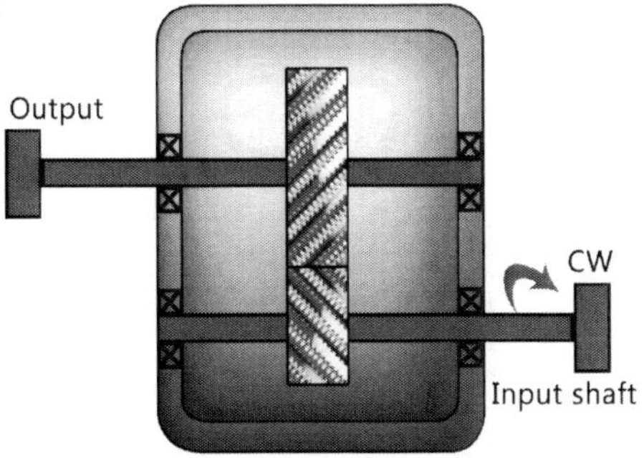

Fault Analysis and Correction

Question 122

Study the next simplified illustration for a gearbox. Input speed is 1500 rpm. Which statement is true about mass unbalance?

A. The unbalance frequency of the input shaft > 1500 cpm

B. The unbalance frequency of the input shaft < 1500 cpm

C. The unbalance frequency of the input shaft = 1500 cpm

D. The unbalance frequency of the output shaft = 1500 cpm

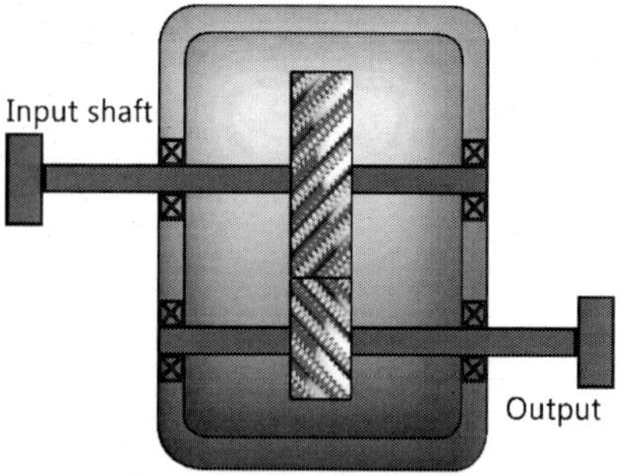

Fault Analysis and Correction

Question 123

Vibration data was collected from a small pump driven by an electric motor (3580 rpm). The single stage centrifugal pump (7 vanes) is connected to the motor by a flexible coupling. The vibration spectrum shows a strong 3X while other frequencies show small amplitudes. What is the most probable cause of this vibration? Hint: Identify the order of each of the given faults.

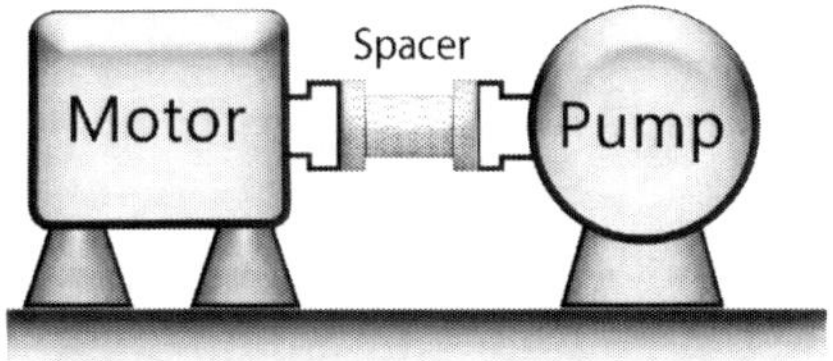

A. Coupling issue like shaft misalignment

B. Cavitation at the inlet of the pump

C. Electrical problem like over load

D. Mass unbalance on the coupling or impeller

Note : ....

Fault Analysis and Correction

Question 124

What is the GMF of the following gearbox? There are 43 and 124 teeth in the inut and output gears, respectively.

A. Number of teeth x 60 cpm

B. Number of teeth x 60 Hz

C. GMF cannot be found because the diameters of the gears are needed.

D. GMF cannot be found because the speed of the shaft is needed.

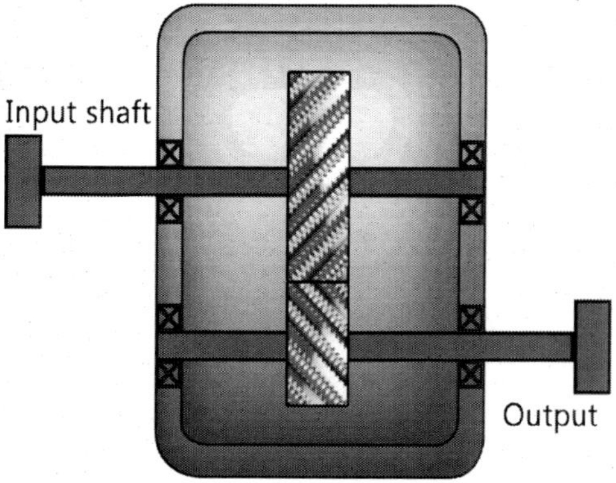

Fault Analysis and Correction

Question 125

Consider a horizontal centrifugal fan. What is a key vibration symptom of a rotor unbalance?

A. In waveforms, there is an intermittent spike component.

B. In frequency spectra, there is a major component at 2 times running speed.

C. In frequency spectra, there is a major component at running speed.

D. A and C

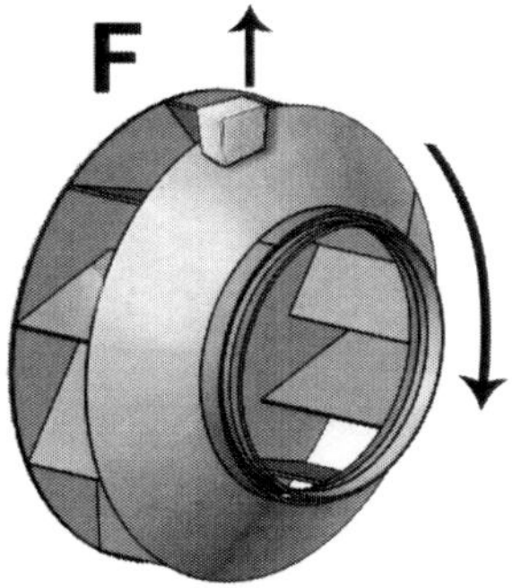

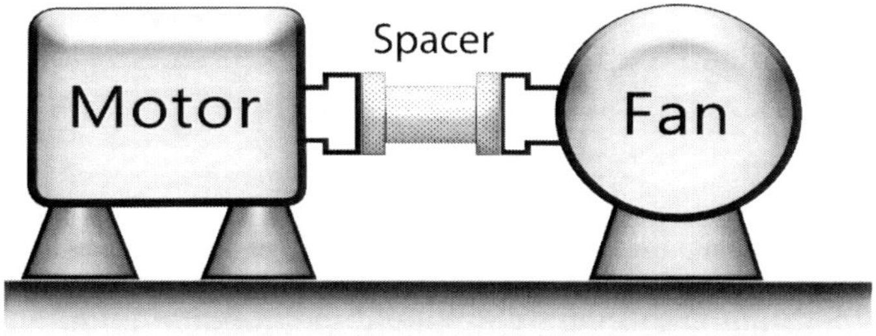

Fault Analysis and Correction

Question 126

Consider a horizontal centrifugal blower. What is the strongest symptom of looseness problem that can be found in vibration plots? Select the best choice from the given answers.

A. High subharmonic amplitudes

B. High harmonics of running speed

C. High 2LF regardless of the load

D. A large component at 7200 Hz if the line frequency is 60 Hz (or 6000 Hz if the frequency is 50 Hz)

Note : ....

....

....

Fault Analysis and Correction

Question 127

Consider a small horizontal electric motor. What can you find in a vibration a spectrum with an excessive looseness of a bearing in its housing?

A. High subharmonic amplitudes

B. A combination of 1X, 2X, 3X and 4X amplitudes

C. High GMF and 3X regardless of bearing size

D. High amplitudes at bearing fault frequency (both inner race and outer race)

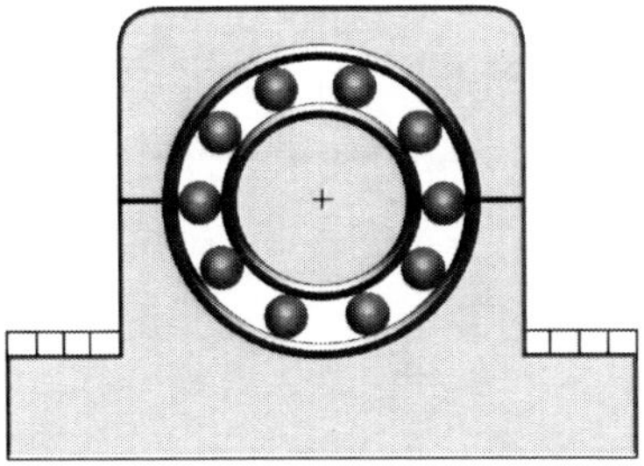

Fault Analysis and Correction

Question 128

What is the most indicative symptom of a structural resonance of a rotating equipment like a vertical pump? Select the best answer from the given choices.

A. In frequency spectrum, there is a major component at 3.33 X times running speed.

B. In a waveform, there are three (3) clear spikes at running speed.

C. Unstable amplitude at 3X clearly seen in the waveform.

D. Abnormally high vibration (usually at 1X) in one direction compared to much lower amplitude on the other direction.

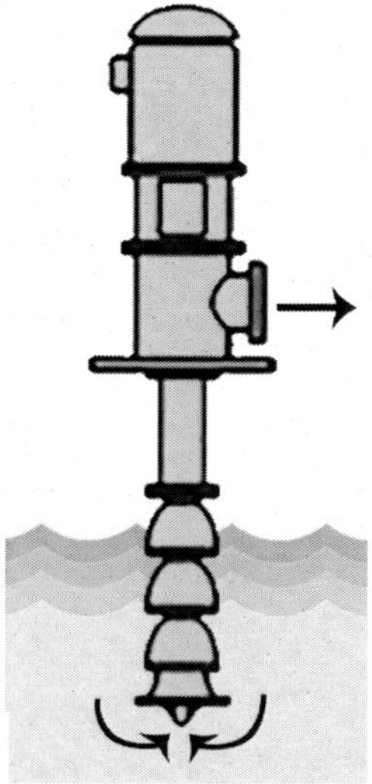

Fault Analysis and Correction

Question 129

What is a key vibration symptom that indicates a possible shaft misalignment problem? Consider a typical horizontal medium size machine. Select the best answer from the given choices.

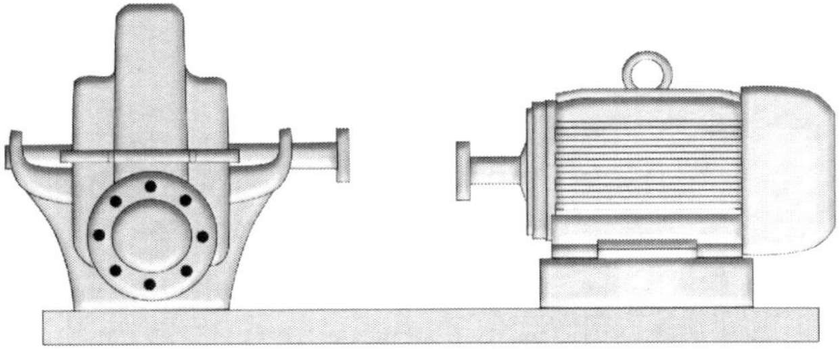

A. 1X, 2X and/or 3X radial vibration, sometimes accompanied by axial vibration.

B. High frequency vibration (very clear when measured in acceleration).

C. High amplitude at 120 Hz measured near to the coupling.

D. All of the above.

Fault Analysis and Correction

Question 130

What is a key vibration symptom that indicates a possible electric fault of an induction motor?

A. Usually, the appearance of 2LF (for example 120 Hz for a 60 Hz electric power).

B. Sudden vibration drop once the electric power is cut.

C. High floor noise on the spectra collected on all directions.

D. A and B

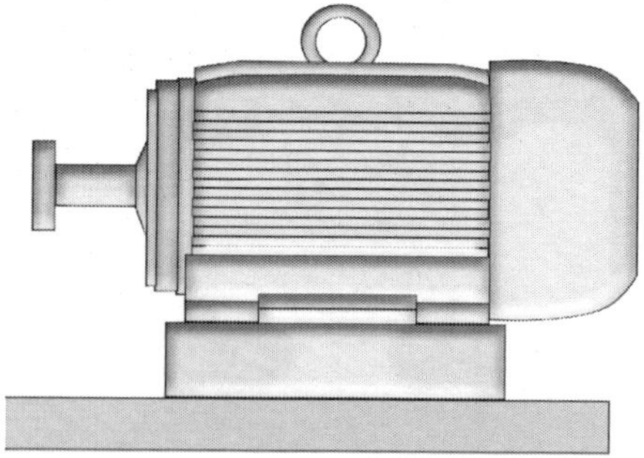

Fault Analysis and Correction

Question 131

What is a key vibration symptom that indicates an operational (process) problem of a centrifugal pump?

A. Audible noise and usually unstable vibration.

B. Vane pass frequency and an elevated floor noise in the spectrum.

C. Harmonics of 1X and a 360 degree phase difference between horizontal and vertical directions.

D. A and B.

Fault Analysis and Correction

Question 132

What is a key vibration symptom that indicates a possible operational upset of a centrifugal fan?

A. Audible noise and usually unstable vibration.

B. Vane pass frequency and an elevated floor noise in the spectrum.

C. Harmonics of 1X and a 360 degree phase difference between horizontal and vertical directions.

D. A and B.

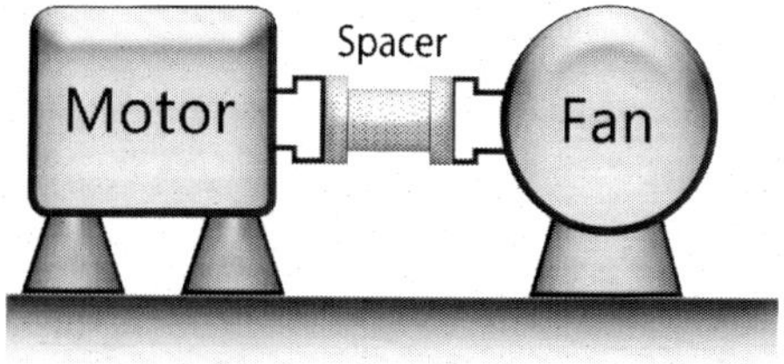

Fault Analysis and Correction

Question 133

Consider the next single stage (between bearing) pump driven by a motor (not shown). What is the difference between a typical shaft misalignment and a typical rotor unbalance when you analyze the vibration spectra?

	Unbalance	Misalignment
A.	High 1X	High 1X, 2X or 3X
B.	High radial amplitude	High radial and/or axial amplitude
C.	Very sensitive to speed	Less sensitive to speed
D.	All of the above	

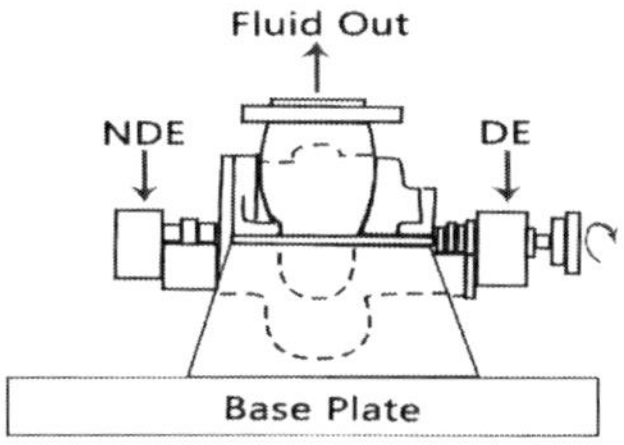

Fault Analysis and Correction

Question 134

A vertical pump is driven by a four pole motor (1790 rpm). The unit has a natural the frequency of 1812 cpm which is very close to the shaft's rotational speed. The pump vibrates severely at shaft running speed (1X) even after the motor was balanced in the shop. What could be the most probable cause of this vibration? Refer to the next illustration.

A. Unbalance

B. Resonance

C. Electrical problem

D. High suction pressure

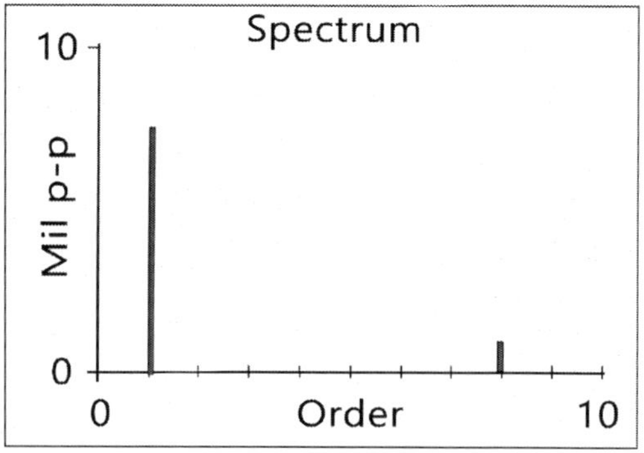

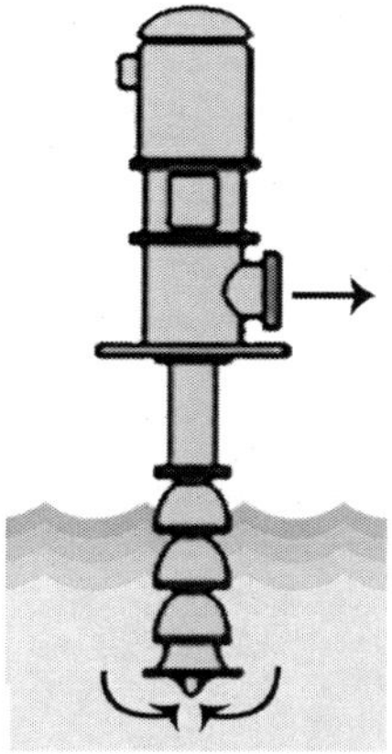

Fault Analysis and Correction
Question 135

Consider a typical horizontal blower driven by a motor. What is the difference between shaft misalignment and part looseness when you analyze the vibration spectra?

	Misalignment	Looseness
A.	1X, 2X, 3X or a combination of them	1X and harmonics
B.	Stable waveform	Could be an erratic waveform
C.	Amplitude and phase are consistent after each startup	Amplitude and phase may vary after each startup
D.	All of the above	

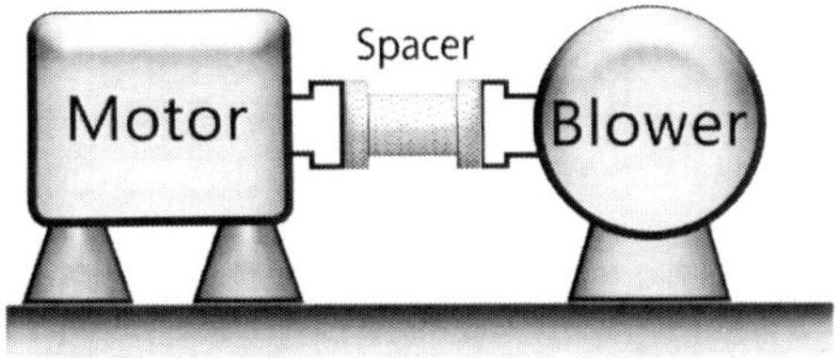

Fault Analysis and Correction

Question 136

This spectrum is collected from the NDE horizontal direction of a vertical motor driving a pump. The vibration amplitude on the same bearing but 90 degree apart shows very low amplitude. Gradual vibration increase is noticed from the past year trend. What is the most possible fault based on the next spectrum if the pump operates at 2375 rpm?

A. Shaft misalignment

B. Resonance

C. Aliasing

D. Varnishing

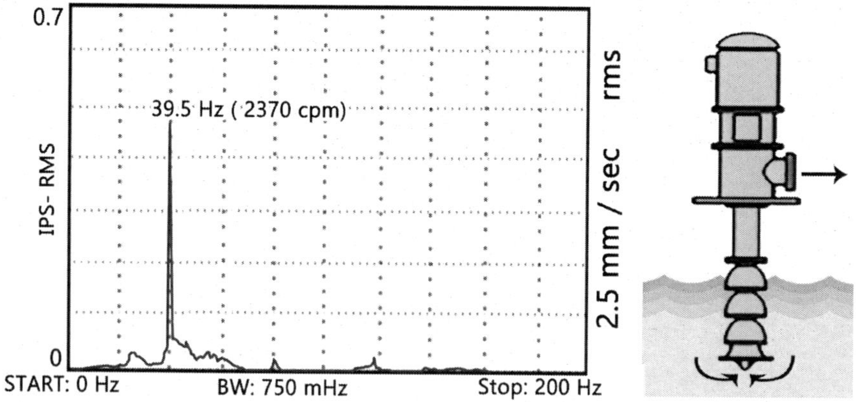

Fault Analysis and Correction

Question 137

A motor runs with high vibration amplitudes when measured in the horizontal and vertical directions. How can an analyst confirm that it is a resonance case not an unbalance?
By comparing the ______________.

A. frequency of 1X with frequency of 2X.

B. natural frequency with the vibration frequency.

C. vibration frequency with stiffness of the machine.

D. vibration frequency with mass (weight) of the machine.

Note : ....

....

....

Fault Analysis and Correction

Question 138

As rotating equipment develop a resonance condition at 1X, the vibration amplitude suddenly drops. True or False?

A. True. This is a classic symptom of resonance.

B. False. Amplitude would likely increase at resonance.

C. False. In fact, frequency (F_{max}) drops at resonance.

D. False. Vibration suddenly drops at cavitation (not at resonance).

Note : ....

....

....

Fault Analysis and Correction
Question 139

What is the most probable frequency of vibration for the following case? Vibration suddenly increased on a vertical motor (driving a sea water pump) after a sand storm. The impeller of the pump has four vanes.

Shaft speed	= 1495 rpm	Line Frequency	= 3000 cpm
Natural Frequency	= 1450 cpm	2LF	= 6000 cpm
2X	= 2990 cpm	Top Bearing Cage Pass	= 585 cpm

A. 585 cpm

B. 1495 cpm

C. 5980 cpm

D. 6000 cpm

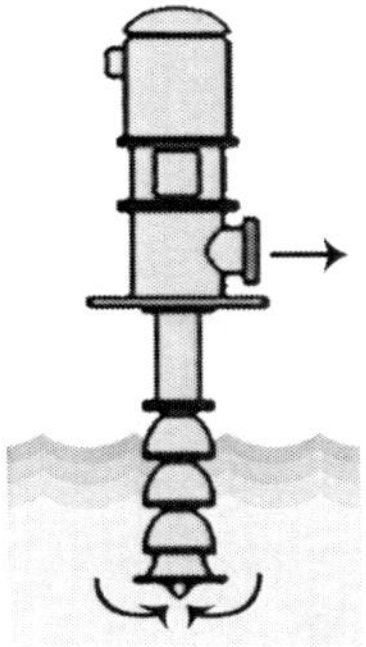

Fault Analysis and Correction

Question 140

What is the most probable vibration frequency of for the following case? High vibration is recorded on the top bearing of a vertical motor. The pump delivers cooling water from a dirty lake.

Shaft speed	= **1795** rpm	Electrical Line Frequency	= **3600** cpm
Natural Frequency	= **1750** cpm	2LF	= **7200** cpm
Vanes	= 3	Resolution	= 10 cpm

A. 1795 cpm

B. 3590 cpm

C. 5385 cpm

D. 7200 cpm

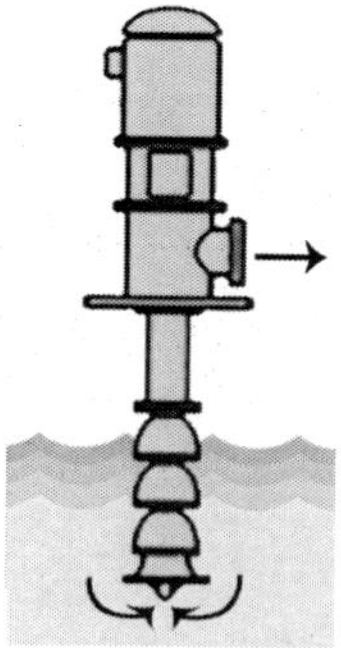

Fault Analysis and Correction

Question 141

Consider a high resolution spectrum of a centrifugal pump (whose shaft is mounted on ball bearings). The vibration spectrum shows a low 1X amplitude but a high 5X amplitude. What could be the most probable problem? Select from the given choices Hint: Sketch the spectrum.

A. Bearing defect

B. Shaft misalignment

C. Hydraulic problem like cavitation or recirculation

D. Coupling unbalance

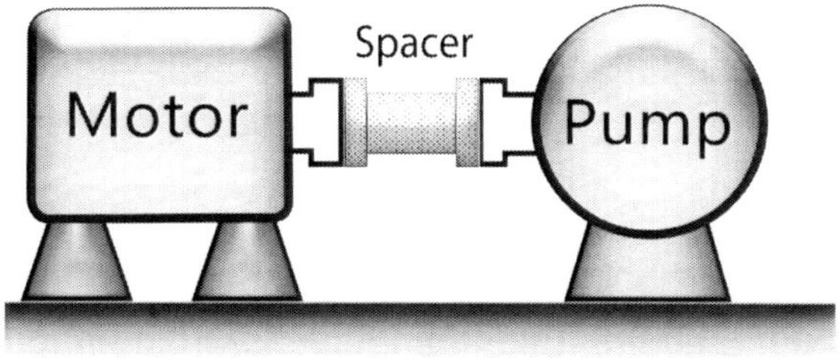

Fault Analysis and Correction

Question 142

This spectrum is for a small horizontal centrifugal pump which is driven by an induction motor running at 1785 rpm. The impeller has several vanes and the bearings are anti-friction type. What are the most probable cause or causes of the high vibration?

A. Hydraulic problem like cavitation

B. Severe shaft misalignment

C. Rolling element bearing defect

D. A and C

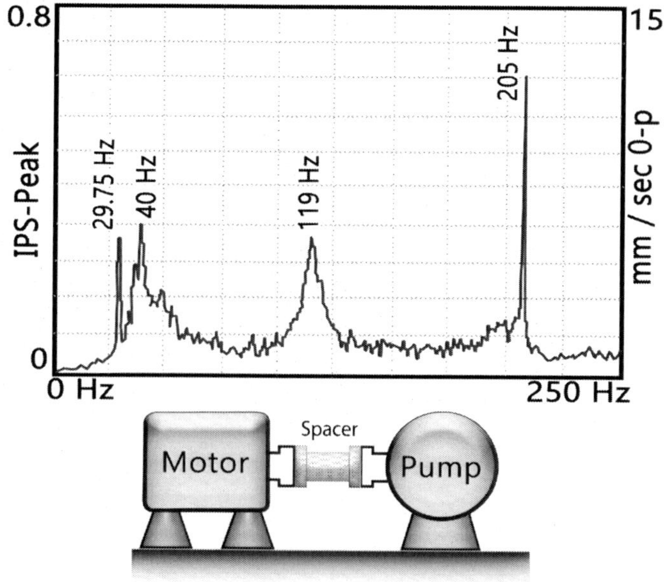

Fault Analysis and Correction

Question 143

Consider a fan (5 vanes) with a large mass imbalance. This fan is mounted on simple sleeve bearings and its speed is 1195 rpm. From the operational (process) parameters, the fan is healthy. At what frequency order would the fan vibrate due to the explained abnormality?

A. 1/5 X

B. 1X

C. 5975 cpm

D. Vane pass frequency or 5X

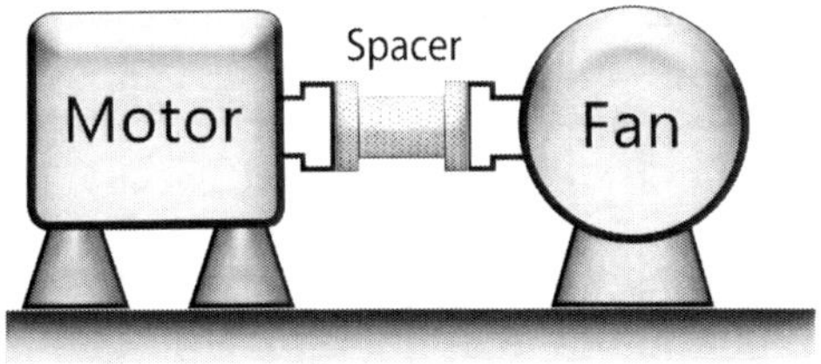

Note : ....

....

....

Fault Analysis and Correction

Question 144

This spectrum is collected from the horizontal direction on the DE bearing of a horizontal pump. The vertical and axial vibrations on the inboard bearing show very low amplitudes. What is the most possible fault based on the next spectrum if this pump operates at 1190 rpm?

A. Misalignment

B. Impeller mass unbalance

C. Cavitation

D. Faulty sensor

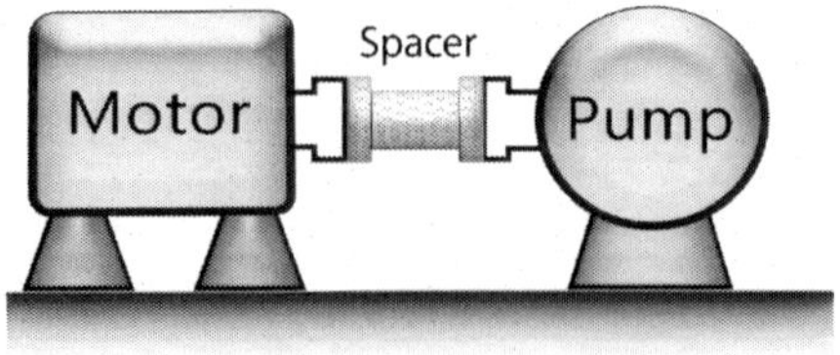

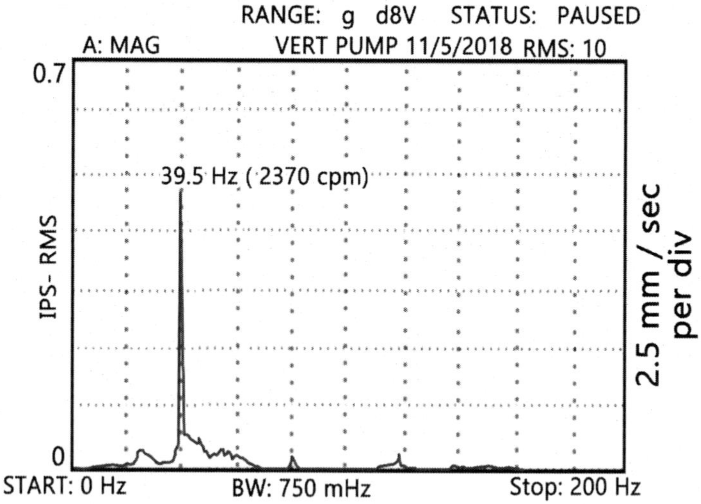

Fault Analysis and Correction

Question 145

This spectrum is collected from a seawater centrifugal pump on the inboard bearing at the horizontal direction. The vertical direction shows a similar signature. What is the most possible fault based on the spectrum if the pump operates at 474 rpm?

A. Rotor unbalance (dirty impeller)

B. Hydraulic problem (vane pass frequency)

C. Looseness (parts with excessive clearance)

D. Faulty sensor (looseness in wires)

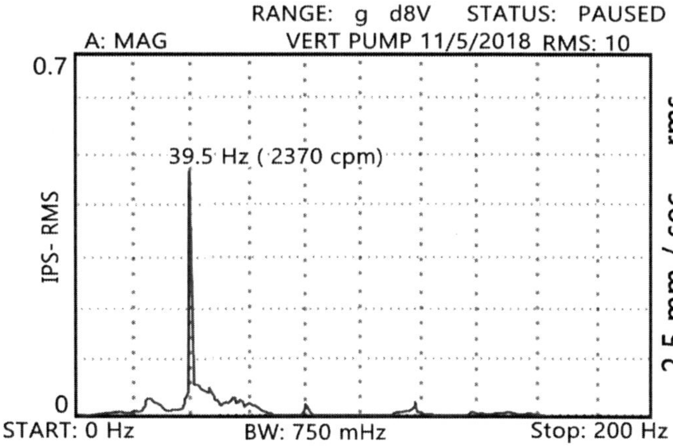

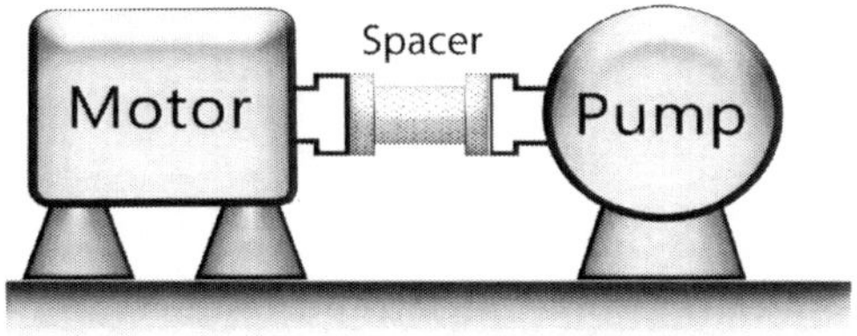

Fault Analysis and Correction

Question 146

The shown vibration spectrum is collected from the NDE bearing housing of a fan running at 3375 rpm (56.25 Hz). No operational abnormality is observed.

What is the most probable fault from this vibration spectrum?

A. Rotor unbalance

B. Defected vibration sensor

C. Antifriction bearing defect

D. Looseness (e.g. bearing housing)

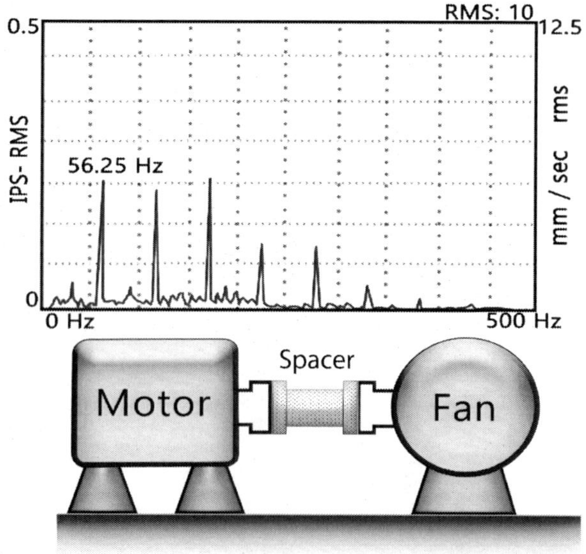

Fault Analysis and Correction

Question 147

Consider a slightly high vibration signal from a bearing housing of a small industrial fan. This fan takes air from a dusty tank. The spectrum shows several peaks at non-integer orders of running speed and some floor noise. What a corrective action should the vibration analyst consider as the first choice for this case? Select from the given options.

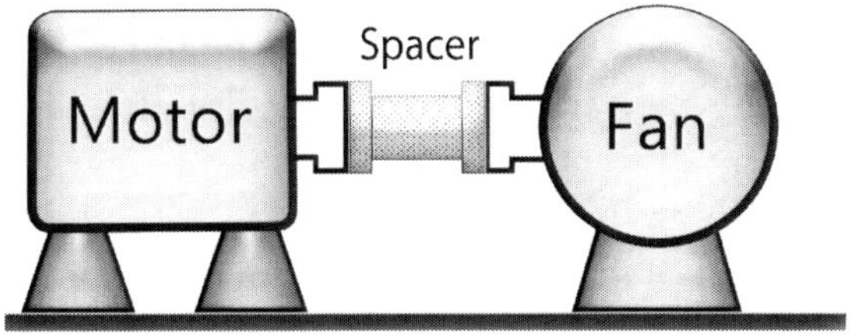

A. Lubricate the bearings

B. Balance the fan's rotor

C. Align shafts

D. Replace the sensor and take another reading

Fault Analysis and Correction

Question 148

Consider a vibration signal from a bearing housing of a small industrial fan (between bearing). The signature shows a low 1x but a high 2X component. What a corrective action should the vibration analyst consider for this case? Select from the given choices.

A. Measure and correct shaft misalignment

B. Lubricate the bearings

C. Balance the fan's rotor

D. Replace the cooling fan of the motor

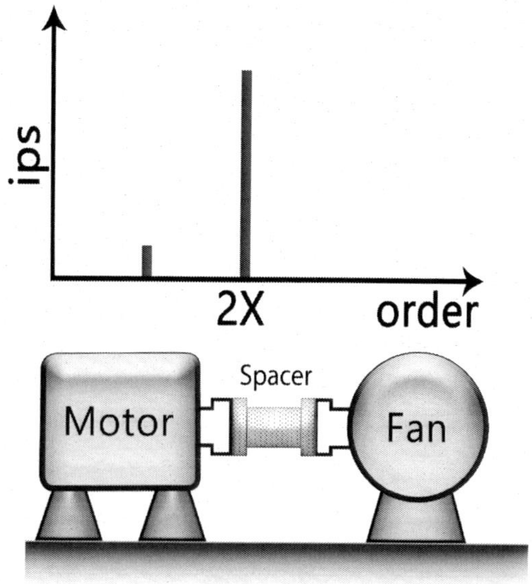

Fault Analysis and Correction
Question 149

Vibration signitures were collected from a bearing housing of a small horizontal fan. The reading indicates 1X and 2X components in both horizontal and vertical directions. The fan has recently been installed after it was perfectly repaired in a shop.

What is the most suitable corrective action the vibration needs to consider based on the most probable vibration cause for this case? Select the best from the given choices.

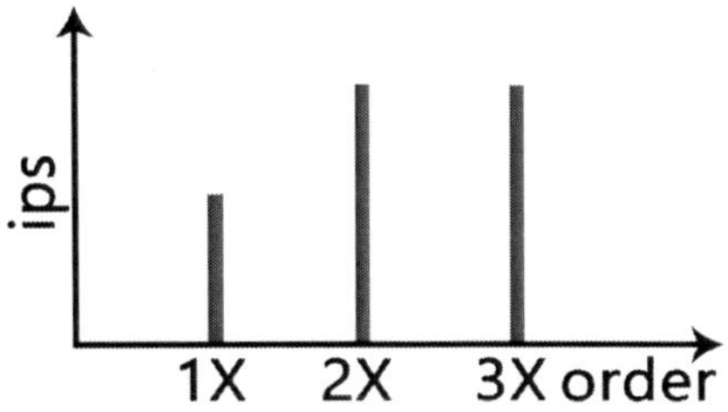

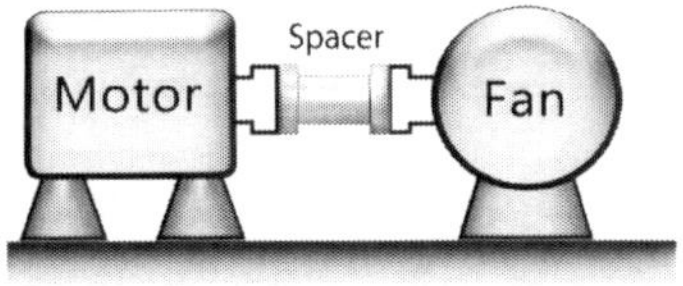

A. Inspect coupling and improve shaft alignment

B. Replace the bearings

C. Take the fan to the shop for balancing checks

D. Clean inlet filters, open outlet valve and perform a visual inspection for the impeller

Fault Analysis and Correction

Question 150

What is true about "beat" condition when you use this term in the field of machinery vibration analysis? Hint: Sketch a simplified signal for beat.

A. A condition at which vibration phase angles are equal in both X and Y directions.

B. Cyclic amplitude variation observed in the overall trend and waveform plots.

C. Very high amplitude at a high frequency above the approved shut down set points.

D. A condition at which vibration amplitudes equal to the bearing clearance for journal bearings.

Fault Analysis and Correction

Question 151

In the field of vibration analysis, which of the following vibration plots is the best to indicate a frequency modulation pattern.

A. 1

B. 2

C. 3

D. 4

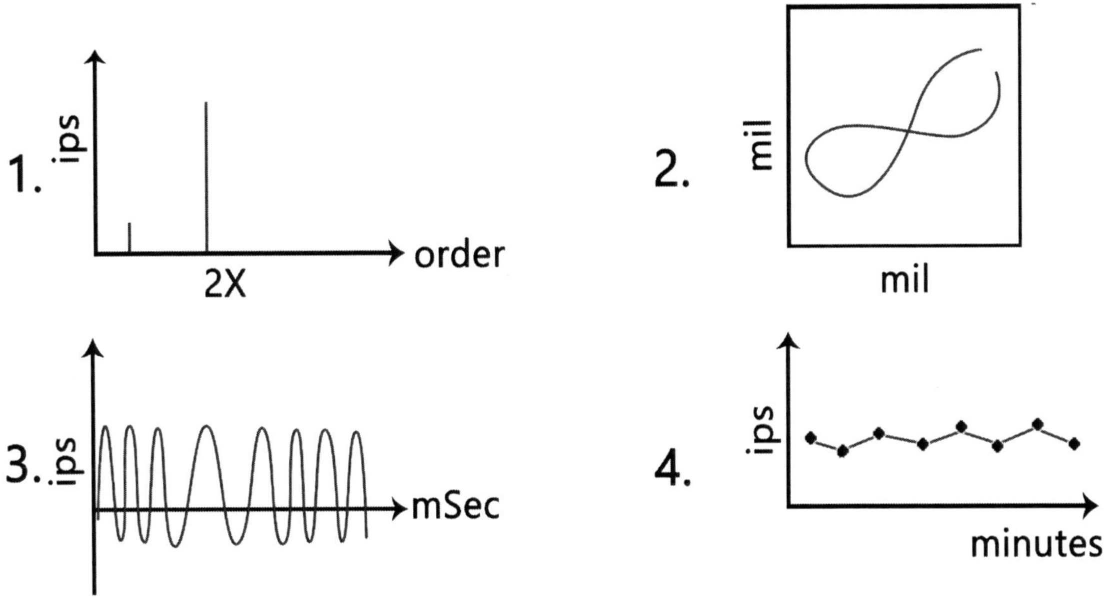

Fault Analysis and Correction

Question 152

In the field of vibration analysis, which of the following vibration plots is the best to indicate a possible beat?

A. 1

B. 2

C. 3

D. 4

1.

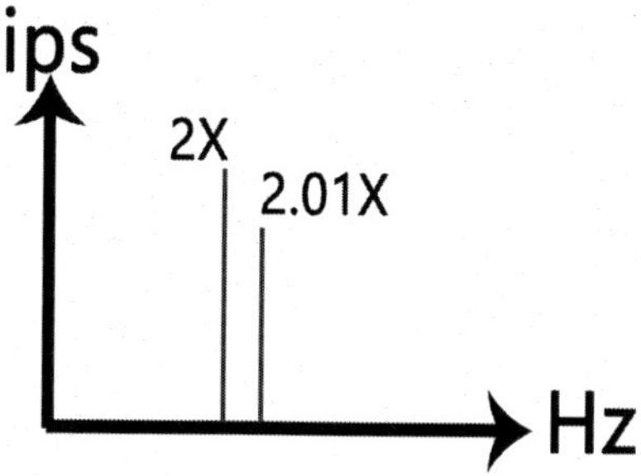

2.

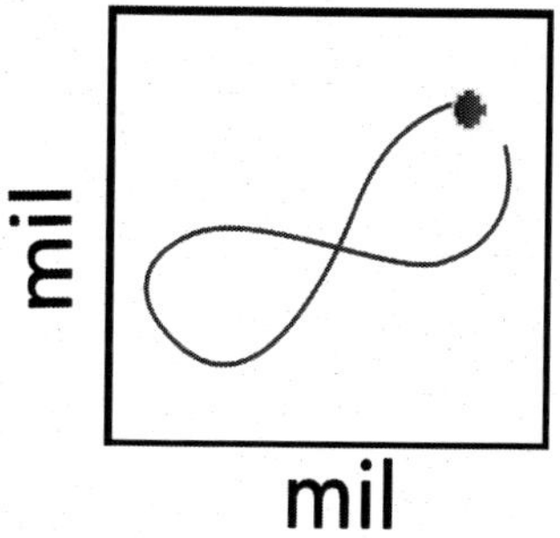

3.

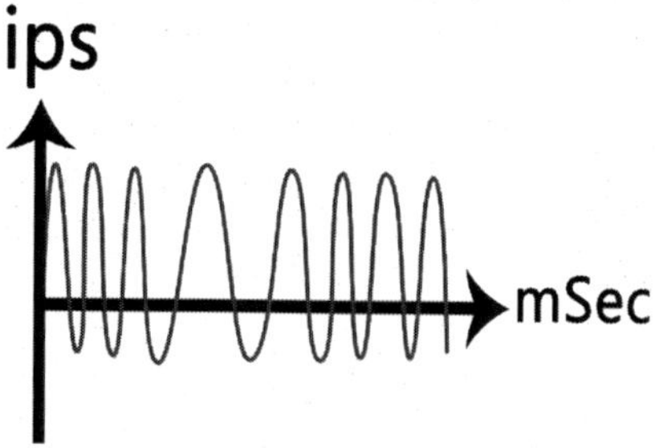

4.

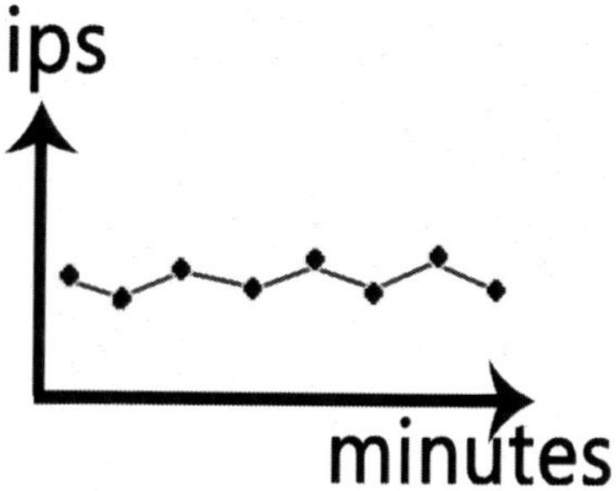

Fault Analysis and Correction

Question 153

Study the following waveforms. Which vibration plot is the best to illustrate a pattern of "beat"?

A. 1

B. 2

C. 3

D. 4

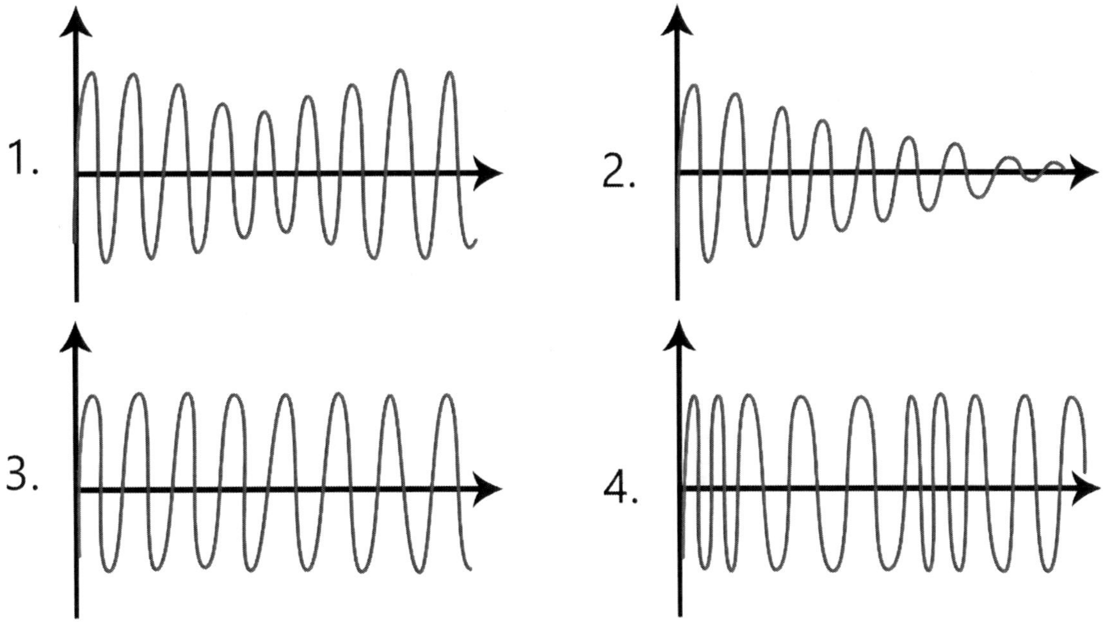

Fault Analysis and Correction

Question 154

Which of the following vibration waveforms demonstrates a possible machine trip?

A. 1

B. 2

C. 3

D. 4

1. 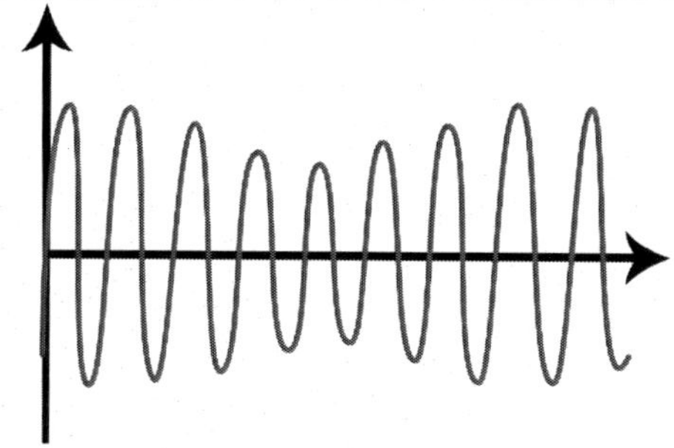

2. 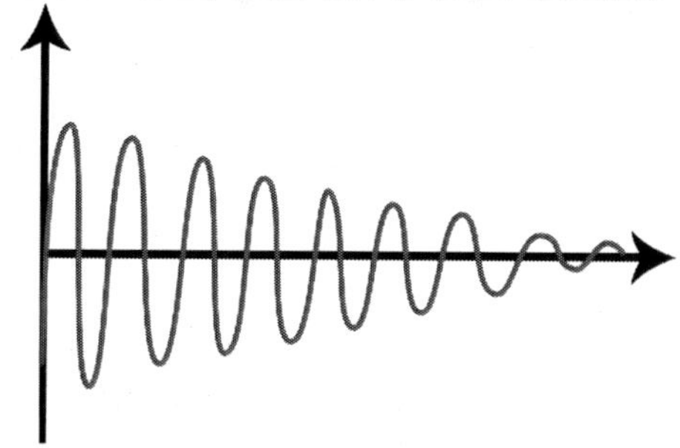

3. 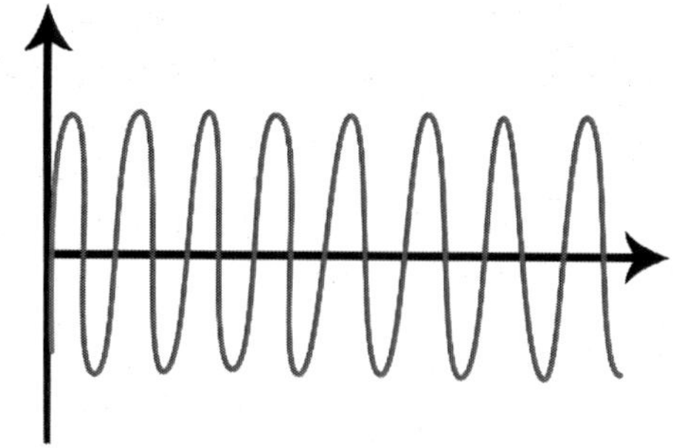

4. 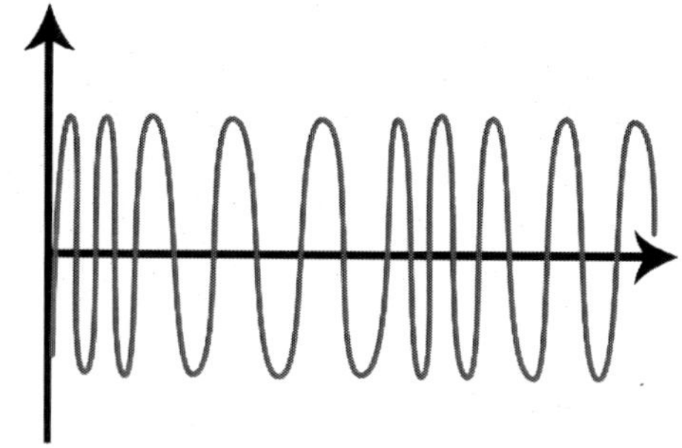

Notes

Notes

ANSWER SHEETS

EMPTY ANSWER SHEET 1

1	A	B	C	D
2	A	B	C	D
3	A	B	C	D
4	A	B	C	D
5	A	B	C	D
6	A	B	C	D
7	A	B	C	D
8	A	B	C	D
9	A	B	C	D
10	A	B	C	D
11	A	B	C	D
12	A	B	C	D
13	A	B	C	D
14	A	B	C	D
15	A	B	C	D

16	A	B	C	D
17	A	B	C	D
18	A	B	C	D
19	A	B	C	D
20	A	B	C	D
21	A	B	C	D
22	A	B	C	D
23	A	B	C	D
24	A	B	C	D
25	A	B	C	D
26	A	B	C	D
27	A	B	C	D
28	A	B	C	D
29	A	B	C	D
30	A	B	C	D

EMPTY ANSWER SHEET 1

31	A	B	C	D
32	A	B	C	D
33	A	B	C	D
34	A	B	C	D
35	A	B	C	D
36	A	B	C	D
37	A	B	C	D
38	A	B	C	D
39	A	B	C	D
40	A	B	C	D
41	A	B	C	D
42	A	B	C	D
43	A	B	C	D
44	A	B	C	D
45	A	B	C	D

46	A	B	C	D
47	A	B	C	D
48	A	B	C	D
49	A	B	C	D
50	A	B	C	D
51	A	B	C	D
52	A	B	C	D
53	A	B	C	D
54	A	B	C	D
55	A	B	C	D
56	A	B	C	D
57	A	B	C	D
58	A	B	C	D
59	A	B	C	D
60	A	B	C	D

EMPTY ANSWER SHEET 1

61	A	B	C	D
62	A	B	C	D
63	A	B	C	D
64	A	B	C	D
65	A	B	C	D
66	A	B	C	D
67	A	B	C	D
68	A	B	C	D
69	A	B	C	D
70	A	B	C	D
71	A	B	C	D
72	A	B	C	D
73	A	B	C	D
74	A	B	C	D
75	A	B	C	D

76	A	B	C	D
77	A	B	C	D
78	A	B	C	D
79	A	B	C	D
80	A	B	C	D
81	A	B	C	D
82	A	B	C	D
83	A	B	C	D
84	A	B	C	D
85	A	B	C	D
86	A	B	C	D
87	A	B	C	D
88	A	B	C	D
89	A	B	C	D
90	A	B	C	D

EMPTY ANSWER SHEET 1

91	A	B	C	D
92	A	B	C	D
93	A	B	C	D
94	A	B	C	D
95	A	B	C	D
96	A	B	C	D
97	A	B	C	D
98	A	B	C	D
99	A	B	C	D
100	A	B	C	D
101	A	B	C	D
102	A	B	C	D
103	A	B	C	D
104	A	B	C	D
105	A	B	C	D

106	A	B	C	D
107	A	B	C	D
108	A	B	C	D
109	A	B	C	D
110	A	B	C	D
111	A	B	C	D
112	A	B	C	D
113	A	B	C	D
114	A	B	C	D
115	A	B	C	D
116	A	B	C	D
117	A	B	C	D
118	A	B	C	D
119	A	B	C	D
120	A	B	C	D

EMPTY ANSWER SHEET 1

121	A	B	C	D
122	A	B	C	D
123	A	B	C	D
124	A	B	C	D
125	A	B	C	D
126	A	B	C	D
127	A	B	C	D
128	A	B	C	D
129	A	B	C	D
130	A	B	C	D
131	A	B	C	D
132	A	B	C	D
133	A	B	C	D
134	A	B	C	D
135	A	B	C	D

136	A	B	C	D
137	A	B	C	D
138	A	B	C	D
139	A	B	C	D
140	A	B	C	D
141	A	B	C	D
142	A	B	C	D
143	A	B	C	D
144	A	B	C	D
145	A	B	C	D
146	A	B	C	D
147	A	B	C	D
148	A	B	C	D
149	A	B	C	D
150	A	B	C	D

EMPTY ANSWER SHEET 1

151	A	B	C	D
152	A	B	C	D
153	A	B	C	D
154	A	B	C	D

Notes

Notes

EMPTY ANSWER SHEET 2

1	A	B	C	D
2	A	B	C	D
3	A	B	C	D
4	A	B	C	D
5	A	B	C	D
6	A	B	C	D
7	A	B	C	D
8	A	B	C	D
9	A	B	C	D
10	A	B	C	D
11	A	B	C	D
12	A	B	C	D
13	A	B	C	D
14	A	B	C	D
15	A	B	C	D

16	A	B	C	D
17	A	B	C	D
18	A	B	C	D
19	A	B	C	D
20	A	B	C	D
21	A	B	C	D
22	A	B	C	D
23	A	B	C	D
24	A	B	C	D
25	A	B	C	D
26	A	B	C	D
27	A	B	C	D
28	A	B	C	D
29	A	B	C	D
30	A	B	C	D

EMPTY ANSWER SHEET 2

31	A	B	C	D
32	A	B	C	D
33	A	B	C	D
34	A	B	C	D
35	A	B	C	D
36	A	B	C	D
37	A	B	C	D
38	A	B	C	D
39	A	B	C	D
40	A	B	C	D
41	A	B	C	D
42	A	B	C	D
43	A	B	C	D
44	A	B	C	D
45	A	B	C	D

46	A	B	C	D
47	A	B	C	D
48	A	B	C	D
49	A	B	C	D
50	A	B	C	D
51	A	B	C	D
52	A	B	C	D
53	A	B	C	D
54	A	B	C	D
55	A	B	C	D
56	A	B	C	D
57	A	B	C	D
58	A	B	C	D
59	A	B	C	D
60	A	B	C	D

EMPTY ANSWER SHEET 2

61	A	B	C	D
62	A	B	C	D
63	A	B	C	D
64	A	B	C	D
65	A	B	C	D
66	A	B	C	D
67	A	B	C	D
68	A	B	C	D
69	A	B	C	D
70	A	B	C	D
71	A	B	C	D
72	A	B	C	D
73	A	B	C	D
74	A	B	C	D
75	A	B	C	D

76	A	B	C	D
77	A	B	C	D
78	A	B	C	D
79	A	B	C	D
80	A	B	C	D
81	A	B	C	D
82	A	B	C	D
83	A	B	C	D
84	A	B	C	D
85	A	B	C	D
86	A	B	C	D
87	A	B	C	D
88	A	B	C	D
89	A	B	C	D
90	A	B	C	D

EMPTY ANSWER SHEET 2

91	A	B	C	D
92	A	B	C	D
93	A	B	C	D
94	A	B	C	D
95	A	B	C	D
96	A	B	C	D
97	A	B	C	D
98	A	B	C	D
99	A	B	C	D
100	A	B	C	D
101	A	B	C	D
102	A	B	C	D
103	A	B	C	D
104	A	B	C	D
105	A	B	C	D

106	A	B	C	D
107	A	B	C	D
108	A	B	C	D
109	A	B	C	D
110	A	B	C	D
111	A	B	C	D
112	A	B	C	D
113	A	B	C	D
114	A	B	C	D
115	A	B	C	D
116	A	B	C	D
117	A	B	C	D
118	A	B	C	D
119	A	B	C	D
120	A	B	C	D

EMPTY ANSWER SHEET 2

121	A	B	C	D
122	A	B	C	D
123	A	B	C	D
124	A	B	C	D
125	A	B	C	D
126	A	B	C	D
127	A	B	C	D
128	A	B	C	D
129	A	B	C	D
130	A	B	C	D
131	A	B	C	D
132	A	B	C	D
133	A	B	C	D
134	A	B	C	D
135	A	B	C	D

136	A	B	C	D
137	A	B	C	D
138	A	B	C	D
139	A	B	C	D
140	A	B	C	D
141	A	B	C	D
142	A	B	C	D
143	A	B	C	D
144	A	B	C	D
145	A	B	C	D
146	A	B	C	D
147	A	B	C	D
148	A	B	C	D
149	A	B	C	D
150	A	B	C	D

EMPTY ANSWER SHEET 2

151	A	B	C	D
152	A	B	C	D
153	A	B	C	D
154	A	B	C	D

Notes

Notes

ANSWER KEY

Q. #	Answer
1	A
2	D
3	D
4	C
5	D
6	C
7	C
8	A
9	C
10	D
11	D
12	D
13	D
14	B
15	A

Q. #	Answer
16	C
17	B
18	A
19	D
20	B
21	A
22	B
23	B
24	A
25	B
26	D
27	B
28	A
29	A
30	C

Q. #	Answer
31	D
32	D
33	C
34	A
35	B
36	B
37	A
38	A
39	B
40	B
41	B
42	A
43	D
44	D
45	B

Q. #	Answer
46	A
47	D
48	C
49	C
50	D
51	C
52	D
53	A
54	D
55	D
56	C
57	B
58	C
59	A
60	D

ANSWER KEY

Q. #	Answer
61	C
62	B
63	B
64	D
65	A
66	B
67	A
68	D
69	C
70	D
71	B
72	C
73	C
74	D
75	C

Q. #	Answer
76	C
77	D
78	D
79	A
80	C
81	D
82	B
83	C
84	A
85	A
86	A
87	D
88	C
89	C
90	A

Q. #	Answer
91	A
92	D
93	D
94	A
95	D
96	B
97	C
98	A
99	A
100	A
101	B
102	B
103	D
104	D
105	D

Q. #	Answer
106	D
107	D
108	D
109	D
110	A
111	B
112	C
113	D
114	A
115	A
116	D
117	D
118	C
119	A
120	D

ANSWER KEY

Q. #	Answer
121	B
122	C
123	A
124	D
125	C
126	B
127	B
128	D
129	A
130	D
131	D
132	D
133	D
134	B
135	D

Q. #	Answer
136	B
137	B
138	B
139	B
140	A
141	C
142	D
143	B
144	A
145	B
146	D
147	A
148	A
149	A
150	B

Q. #	Answer
151	C
152	A
153	A
154	B

Notes

ORDER OTHER PARTS OF CAT I PREP I PACKAGE

Don't guess where your skill stands; certify it. PrepCertify believes that the best preparation for professional certifications is obtained through practicing well-designed real world problems.

Learn what really matters in real world industry while mastering the Body of Knowledge in the certification standards. Your Cat I Prep I series does that for you. Through PrepCertify, you will achieve your certification in a much shorter time and with a greater value of your time and effort.

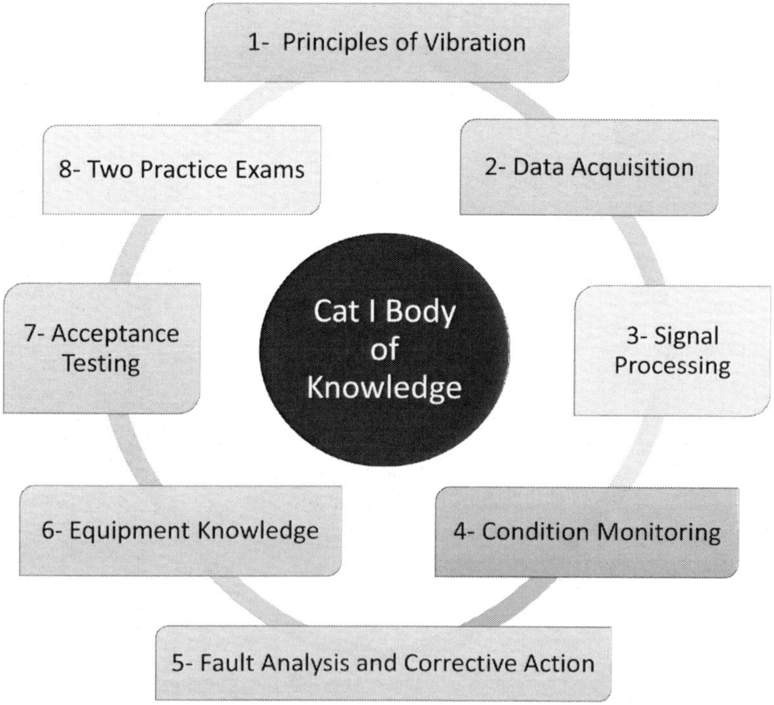

Vibration Analysis Certification Exam Preparation Package
Certified Vibration Analyst Category I
ISO 18436-2 CVA Level 1
CAT I PREP I SERIES PRACTICE TESTS

Part	Covered Body of Knowledge	ISBN-13
1	PRINCIPLES OF VIBRATION	978-1-64415-006-1
2	DATA ACQUISITION	978-1-64415-009-2
3	SIGNAL PROCESSING	978-1-64415-002-3
4	CONDITION MONITORING	978-1-64415-005-4
5	FAULT ANALYSIS AND CORRECTION	978-1-64415-008-5
6	EQUIPMENT KNOWLEDGE	978-1-64415-001-6
7	ACCEPTANCE TESTING	978-1-64415-004-7
8	TWO PRACTICE TESTS	978-1-64415-007-8

Notes

Notes

Made in United States
North Haven, CT
05 November 2026

21766159R00120